Sabra & Bob

A Love Story

Sabra & Bob

A Love Story

Brig. Gen. Robert W. Kennedy USAF (Ret.)

ISBN # 978-1-7343933-7-8

GR8TRANSITIONS4U

Dedication

This book is dedicated to our families who raised us and provided us with our initial set of values and attitudes, to those childhood teachers, friends and neighbors who added value along the way, and to our outstanding friends,supervisors, and comrades-in-arms who shared in our love of God, family and country. Most of all it is dedicated to my wife, Mary Sabra Kennedy, who pro-vided me with the gift of faith by her example. Finally, it is dedicated to all of my comrades-in-arms who were killed or disabled during our 30 years of military service, and Janet Litherland Barnes , my editor, who gave me encouragement during the four years that I struggled writing this book.

This is a love story of a very happy marriage between an undisciplined, curious boy who craved adventure, and a lady with a beautiful soul. It is a happy book written for our children, grandchildren, great-grandchildren and grand-nephews/nieces, and hopefully, it will lead them to happy marriages in an America that is drastically different from the America that we experienced, one that provides much greater challenges to the institution of marriage. My message is simple: A happy marriage translates to a very happy life. It requires a total commitment to each other before God and constant nourishment of your bond. Express your love every day. Never go to bed angry. Stay physically fit. Make your home the neighborhood hangout for your children's friends. Play with them. Eat meals together and talk. Mutually establish goals and objectives and a plan to attain them. Save more than 10 percent of your salary and invest it using dollar-cost averaging. Pass your belief system to your children. Always think positive.

God Speed

About the Author

Brigadier General Bob Kennedy, MSEE, BSEE, MBA served 30 years in the US military. He is a Command Pilot and has been awarded the Distinguished Service Cross, Silver Star, four Distinguished Flying Gross and twelve Air Medals. He served four years in Germany when the Berlin Wall was built, one year in Korea and eight months in Thailand flying in the Korean and Vietnam wars. He spent 27 years in industry following his military service in two executive jobs, two "think tanks" and as an aerospace consultant on more than 200 high value contracts. This is the story of his very happy life with his bride, Mary Sabra Kennedy.

Contents

~ CHAPTER 1 ~

My Family

Mother and Doris

My father, Harry Lyttle Kennedy, emigrated from Northern Ireland in 1922 and married Harriet Louise Williamson on November 27, 1923, in Nolan County, Texas. He was 28 and she was 19. I was their second child.

Harry was employed by the Pauli Company, installing the new electrical systems that opened and closed cell doors in jails. He and my mother traveled extensively, staying several months in each location before moving to the next job. When my sister, Doris, was born on August 7, 1925, they settled in Saint Louis because Doris had a very serious operation to correct a stomach block and was in poor health. Harry continued to travel but worked out of Saint Louis.

Growing Up in the Great Depression

I was born in Saint Louis, December 22, 1929. (Table 1-1, below, shows a comparison of data to provide perspective for the reader.) The depression

was in full swing by this time, and my father lost his job in 1930. The four of us moved into my maternal grandparents' home in Saint Louis. My grandparents had five children, and by 1931 four of the five children had moved back home, along with two of their three grandchildren. My grandfather, Jacob Lindsay Williamson, worked for his brother in the Williamson Outdoor Advertising business and had the only job in the house. My father was unable to find work in Saint Louis and left us in 1933 to search for a job elsewhere. We never saw him again, and after a few letters never heard from him again. I finally found him in 1959 in his hometown of Belfast, Northern Ireland. He had deserted us and never looked back. (More about his story later.)

Thus, my wonderful grandparents had to rent a larger home to accommodate their two unemployed sons, Dale and Donald Williamson, two daughters separated from their unemployed husbands, and my sister and me.

My grandparents' oldest child, Elizabeth (Bess), was married to Seney Huckstep, a very intelligent and religious gentleman who kept his job at Curley Clothing Company and eventually became the treasurer. They had a beautiful daughter Elizabeth (Bee), who was ten years older than me. They always owned a home in the suburbs, while we rented homes in the west end of Saint Louis. Our Aunt Bess was pure fun and we loved her very much. She stayed married to Seney until he died in 1950, and enjoyed a life of luxury, but she always had an open relationship with a male friend. I was a teenager before I figured out that this was terribly wrong and that it had a terrible impact on Seney and Bee.

Bee lived a very active social life, coming out as a debutante and marrying a High School football hero, Gordon McGann. Gordon and Bee were supported by Seney, and our whole family loved them and longed for their frequent visits. Gordon served in the U.S. Marine Corp during World War II in China but did not work after he returned home—Seney always provided them with a home in the suburbs and a ranch house in the foothills south of Saint Louis. After Seney died they used his extensive savings to purchase a small hotel, a waterfront home and a boat in Fort Lauderdale, Florida. Within five years they had to sell the boat, hotel and home, and Bee and Gordon and their two children moved into a smaller home. Gordon had to work for a living.

Grandma Alice (Momo), Grandpa, and sister, Doris

Many years later we cared for Bee in a nearby nursing home after Gordon and Bess had died. She was estranged from her two children, suffered from agoraphobia for more than thirty years and was dying from emphysema. She poured her heart out to us in her last two years, revealing the shame and misery she felt as a result of her parents' troubled marriage. Her early life in high society with many friends was gradually eroded by her phobia and alcohol. Their dysfunctional family group was the major tragedy in our extended family. As we watched our beautiful cousin grow more depressed in her own dysfunctional family, my sister and I both reinforced our own very strong family bonds.

My childhood story would not be complete without describing our relationship with my grandmother's two sisters, Mamie and Eva Cross. My great-grandmother Cross had three daughters, Alice my grandmother (born 1870), Mary ("Mamie" born 1883) and Eva (born 1885). Great-grandfather Cross worked for the railroad. One evening as he walked home from work after being paid, he was robbed and murdered. Great-grandmother Cross was left

My grandparents' generation in Coulterville, Illinois. Grandpa and my father (with hat) in back row. Bess, Doris, Mom, and Momo, fifth through eighth in front row.

with three young daughters, no income, some small savings and her home. She moved the girls into her bedroom and used the rest of her home as a boarding house. Alice was the oldest and married Jacob Williamson in 1891, when she was 21 and he was 26. They moved away as soon as they could afford a home and started their own family, but they continued to help great-grandmother Cross, Mamie and Eva until Mamie and Eva had jobs. Great-grandmother Cross discouraged Mamie and Eva to marry and pledged them to take care of her in her old age. While this sounds selfish in our era, children in those days were expected to support their parents in their declining years. In reality, there was no other alternative, so the family provided life-cycle mutual support to the best of their abilities. An early death or disability of the principal wage earner was a major catastrophe, which only the strongest families survived.

Mamie worked for Laclede Gas Company in Saint Louis for almost 30 years. She was fired during the depression—just before she was eligible for retirement. She was very hurt by this unjust action and never worked again, becoming the full-time housekeeper for her mother and sister. Eva worked for Monsanto Chemical Company for her whole life, rising from bookkeeper to become the president's secretary. She retired at 65 after many years in the executive suite and, after a few months, was brought back to work on full salary and retirement. Mr. Sweeney, her boss, had her picked up in a limousine and driven to and from work.

Just as she was about to retire again, they started to design and program an automated financial information system, and they kept her working until she was 75. She had kept the books in most of the departments and become an administrator to Mr. Sweeney who eventually became President of Monsanto. She probably had the best knowledge of the old manual financial system and was best equipped to describe the requirements to the automated system programmers. She used her influence to help my aunt Gladys to be hired as a dishwasher in the company's cafeteria during the depression. Gladys eventually worked her way up to become the manager of all of the Monsanto food facilities in the Saint Louis plant.

Young Bob Kennedy in downtown St. Louis

All of the Cross women were very determined and productive and earned their way into management, though they were never fairly compensated. My Great-aunt Mamie was unjustly fired to save Laclede Gas Company a meager retirement check. Doris and I received unconditional love from our aunts and great aunts. They doted on our every action and exaggerated our every deed. I am certain that I behaved better because I did not want to disappoint the doting females in our family. They established high expectations and praised our every accomplishment. They also ignored all of my shortcomings. I felt as though the family honor was on my shoulders after my grandfather died. Although I was the only responsible male in the house I was surrounded by strong and determined females.

The job situation steadily improved. Uncle Dale went to work as an electrician in the Granite City Steel Mill in 1933, and Gladys went to work in the cafeteria at Monsanto Chemical Company in 1934. Dale stayed at Granite City Steel until he died and worked his way up to head the electricians union. Donald worked in a variety of jobs, usually as a driver for small neighborhood businesses; and, in the 1950s he and his third wife opened a small grocery store in Wellston and operated it until they retired. My mother did not take a job until I graduated from high school in 1947. She was the principal housekeeper alongside my grandmother and of course was a full-time mother. After I graduated from high school, she went to work at Wesley House, a neighborhood children's recreation agency funded by the Red Feather Charitable organization in St. Louis. She retired from Wesley House in 1970, when she was 66 years old. When she went to work there in 1947, it was an all white, middle-class neighborhood; but it changed during the '50s into an all black, crime-plagued neighborhood, well before she retired in 1970. We never had a car and mother never learned to drive, so she rode the public bus to and from this troubled neighborhood for many years. I never worried about her, because she had known most of the troubled adults when they were children. They loved her and called her "Miss Harriet." I knew she was safe; and she never encountered any danger in her Wesley House. She loved her job, all of the children and all of her colleagues.

Family Structure and Stress

Our family life was similar to many depression-era families in urban areas. The lack of jobs brought many single adults and young families back home to live with their grandparents. This was not unusual for my grandparents' generation (born 1868-1870), since most families in their childhood had three generations in their home and most farm families still had three generations in the home in the 1930s. The three-generation family was (and perhaps still is) the best possible and most natural economic and social approach to life. Children enjoy the luxury of enjoying the unconditional love, teachings and attention of

My Family, 1940
Don, Mother, Gladys, Grandmother, Bess, Grandfather, Dale

their grandparents; and, parents have the support and wisdom of the grandparents to better raise the children and run the household (a full time job prior to the availability of household appliances, automatic heaters and refrigeration units and transportation systems).

The second generation had parents with children and also included the single/widowed adults. Most important, there were several wage earners and a sharing of household costs. Finally, the children had brothers, sisters and cousins to play with and their own little hierarchy with attendant authorities and responsibilities to be earned. It sounds ideal—a socialized economic unit operating in a capitalistic democratic society. It works very well on a family farm, and many new immigrant families still use this structure. It was an economic necessity in our family, but the lack of jobs and the broken marriages made it very tough on all of the adults. The stress was apparent. I must have been very insensitive because I was never aware of the adult stresses until I listened to my mother and Aunt Gladys talk about those days during the last ten years of their lives when we cared for them. I meandered through my childhood, loving every minute of my life within our large extended family, oblivious of most of the family problems other than the dire consequences of my own father's desertion.

~~~

My uncles, Dale and Don, did not perform any household chores nor did they contribute to the household expenses even after they found jobs. They both married two or three times for brief periods but returned home in short order. I remember Don's first wife, a very nice Catholic lady. They had a baby girl, Mary Ann Williamson. Don soon deserted his wife and baby girl and returned home. His wonderful wife raised her daughter by herself in the Catholic school system and Mary Ann was the first Williamson to earn a college degree. She married a gentleman from South America and had many children, but she died young— in her fifties. Although I only knew her briefly, I was very proud of her and her mother.

Both Dale and Don expected my mother to do their laundry and barely tolerated my sister and me. My mother even performed the heavy household chores, such as stoking the coal furnace and carrying out the unburned clinkers to the garbage pit behind the house. I relieved her of that task as soon as I was big enough. After we grew up, Dale and Don became more considerate and treated all of us much better.

My Aunt Gladys was a very strong woman, was very close to my mother, and loved Doris and me. She Was married twice but never had children, so she shared us as a "co-mother." My Grandfather died when I was 11, and Gladys then supported my grandmother, mother, my sister and me. Dale and Don continued to live in the house sporadically but never contributed money for the household.

~~~

My Grandparents loved all four of their grandchildren unconditionally, but my sister and I had a little edge because we always lived with them. Looking back, our living conditions seem grim, but I was blissfully unaware of any negative influences and remember a very happy and contented childhood.

My sister was always sickly and very unhappy and always remembered her childhood and life at home as miserable. I attribute this disparity to the fact that she was six years old when our father deserted her and I was only eight months old. I cannot even remember my father. There is little doubt in my mind that Doris never recovered emotionally after our father deserted us, until she married Phil Jackson.

Our mother and grandmother always spoke proudly of our father and told us that he must be dead or he would have returned home. Also, they never spoke badly about Bess's, Dale's and Don's behavior and emphasized positive thoughts. Gladys was more blunt and rolled her eyes a lot when family matters were described in the usual positive manner. Despite all the positive reinforcement, Doris remained unhappy until she received counseling from a psychologist in her middle teen years and also received understanding and attention from a wonderful and sensitive high school music teacher. Doris found music and became active in the choir. It changed her whole outlook, and in a few years she met and fell in love with a wonderful gentleman, Phil Jackson, who also loved music and was very talented and smart. They were married in 1947, when I was in Alaska, and Phil also became my new male role model.

Doris and Phil, 1947

Role Models

I had a wonderful relationship with my grandfather and we were very close. He was the only positive male role model in my formative years. He was a Bible reader but did not preach to me. He probably should have preached more to Bess, Dale, Don and me, but it just was not his nature. When he died in mid-June 1941 at the age of 77, I was very upset and decided to grow up like him and provide for my family as he did.

My Aunt Bess was married to Seney Huckstep then, who became a positive role model during my teenage years, trying to help me to improve my ways when I went astray. I was headstrong, though, and easily tempted by everything that passed my way. My sister's husband, Phil Jackson, became a very good friend and provided me with intelligent and thoughtful insights, which seemed eventually to sink into my thick head.

Those were wonderful years for me and my family—school teachers, Sunday school teachers and neighborhood friends largely molded my wayward childhood values into a more acceptable set of ethics.

We had a very big advantage over modern-day children: a depressed economy, no television, a full-time mother and grandmother, doting aunts and great aunts, dedicated school teachers in neighborhood schools, a big public park (Forest Park) to play in, affordable electric street cars, lots of good books to read and neighbors who would not hesitate to call your mom or your grandmother if they saw you doing something suspicious.

I also fondly remember the neighborhood policeman who walked our beat. We really respected him. In those days the entire neighborhood helped raise all of the children. Today they would be called busybodies.

I have briefly described our family environment as a background to understanding how I became who I am. It is worthwhile to discuss my perception of the process of how my value system eventually surfaced. My most vivid childhood memories left me with some valid, and some not-so-valid, precepts.

Perceptions

The strongest emotion I felt concerned the emotional damage to my mother's and sister's self-image due to my father's departure and the sorry marital record of all of my aunts and uncles. I could see the strong contrast between my grandfather who stood by his wife and extended family, and my father, Dale and Don, who shifted their responsibilities to my grandfather when things were tough.

I also could see the misery caused by the Great Depression and the lack of jobs. I vowed never to desert my wife and children. This in turn meant that I would have to very carefully select my wife and avoid serious relationships until I was convinced that I truly loved the lady and could provide for her and our family. Then we would marry, have children and never consider divorce as a potential solution to marital or economic problems. I know that a happy marriage like my grandparents enjoyed was a very high priority goal for both my sister and me. We both successfully achieved that goal and enjoyed a very happy married life, which our parents, aunts and uncles failed to achieve.

I often wonder if the children of happily married couples assume that they will have a happy married life and fail to carefully select their mates and work hard on their own marriages. Perhaps the happily married couples need to teach their children why they are happy and how and why they decided to marry. It is clear that our society places less value on marriage today. Many couples lack the legal and spiritual *commitment* required for a happy marriage.

~~~

Many of my neighborhood and childhood friends did not have fathers, and I went to a high school with many classmates who lived in the neighborhood Masonic Orphanage for Boys. The downstream results of the depression, prohibition and the wild "roaring twenties" were all around us, and it was many years before I postulated a correlation of the national environment of the twenties, followed by the depression of the thirties, to the beginning of the disruption of family life and values.

Dissolution of the Three-Generation Family

Then World War II started and drafted all of the young men away, bringing scores of families off the farm and into war-production factories. Similarly, many young women went to work in the factories or into the military. The depression ended with the beginning of World War II, and all of a sudden everybody had a job. However, a massive disruption of family life then occurred, with a whole generation of young people moving away from the farm, all of the young men going to war, and an enormous number of male and female workers going into the factories to support the largest industrial mobilization in history. The three-generation family became very rare.

~~~

When World War II ended in 1945, all of the young men came home, many married their sweethearts and the baby boom began. Housing developments sprung up all over the United States, and the single-family home in a development became the national norm. Many former military personnel used the G.I. Bill of Rights to win their college degrees or attended industrial training courses to qualify for well-paid technical jobs. The G.I. Bill of Rights was a key critical decision that provided long-term growth and prosperity to most Americans.

The 1950s saw rapid growth in our economy, education, employment, many labor-saving home appliances, television and single-generation families.

Unfortunately, the babies produced by the boom had a monetary allowance, more television, less parental (and neighborhood) supervision, and Dr. Spock to replace the wisdom of grandparents. Labor-saving home appliances played a role in allowing more wives to work; many wives already had worked hard in the mobilization industries while their husbands were away at war.

Baby boomers came of age in the middle sixties, and the family would never be the same. The roaring twenties looked mild compared to the selfish and undisciplined "protest" environment of the sixties.

Personal Postulation

I will postulate a strong correlation between the undisciplined sixties and the current divorce rate, abortion rate, illiteracy rate, bankruptcy rates and the substitution of peer-group values for family values. Special-interest organizations now run rampant, seeking to sub-optimize the rights of teachers, women, minorities, lawyers, unions, corporations and various causes using big lies and political influence.

Today, almost everything that is politically correct is factually unsubstantiated. These groups accept and apply the intellectual compromise of the sixties; e.g., the desirability of the "cause" justifies their questionable tactics and lies coupled with the rigorous avoidance of accepting personal responsibility for anything. I attribute much of this to the degradation of our moral and ethical foundations, particularly in the press, entertainment media, advertising, academia and political institutions.

We now live in the information age enabled via high-tech communications and computer aids at affordable prices. Our problem is that we have instant access to many channels of communications, but most of these channels produce trivial content at a fourth-grade level. Our information producers lack the imagination, education and creative talent to provide sufficient quality or reliable information to an under-educated public.

We sell our products, ideas and ethics using politically correct dogma and illogical processes stressing materialistic values. I maintain that this is our primary national problem, and it is directly related to the breakdown of our ethical base in schools, businesses and the family. Only the religious institutions remain standing, and they are being watered down by our courts and legislatures.

I cannot heal this national flaw but only can try to teach my family to obtain the best possible education so that they can support their families, work hard for a good marriage and build upon the religious heritage within their own families. They will have to work hard to pass this on to their descendants in this morally troubled age. It saddens me to have seen the household family unit—the fundamental building block of our country— degrade from three generations per household to one-and-one-half generations per household (single parent with children) in my lifetime. Government (taxpayers) have become responsible for childcare, financial subsidies, sex education, counseling, retirement and care of our elders as each successive generation has shed more and more family responsibilities.

The depression, the information explosion and three wars were the forcing functions behind this social revolution, and I am sure that each generation views the costs and benefits of the resulting changes very differently. My mother would have started this discussion with World War I and prohibition; my grandparents with the Civil War, the gold rush and the railroads; and so on back to the rise and fall of the Roman and Greek empires. All societies present challenges to struggling couples. The entire family—not government—must take responsibility for one another to meet these challenges.

The Melting Pot

Saint Louis was the typical melting pot of many diverse cultures when I was growing up. Various ethnic groups clustered in their own neighborhoods—the Italians in the south side, the Germans in the north and also in the south side, a small Greek community in the west end, and the blacks compressed in an area near the center of the city. There were many three-generation families where the grandparents still spoke their native language, the parents proudly spoke American with an accent and the children spoke straight American plus their native language. Some of the groups distributed a weekly newsletter in their native language, and the older folks would converse in their native language with one another.

My childhood friends always spoke American at the insistence of their parents and grandparents. They had emigrated to America because it offered new and better opportunities and they wanted their children to communicate in their new language and to study hard in school—a stark contrast to today's tribal orientation. We lived in the far west end in a rental neighborhood with a high percentage of Jewish and Irish families. Most of my Jewish and Irish Catholic classmates had fathers and mothers, but most of my friends were from broken families and did not have a father. It was kind of clannish and we were probably the broken family clan. I learned to respect the Jewish culture because Jewish parents stayed together and were very strict on religion. Also, their children studied harder in school and made better grades.

Hellfire, Brimstone and Prejudice

My family was Presbyterian, except for Dale and Don, and I accompanied them to Sunday school and church every Sunday. Our minister was Dr. Lampe and he was not child-friendly. After I "graduated" from Sunday school and started attending the main church services, I was baffled by his fire-and-brimstone approach. According to his sermons, nearly everything I did or wanted to do would lead me straight to hell. Our church seemed to imply that unless you were a very good Presbyterian you were not headed to heaven. I could not correlate the teachings from the bible to the beliefs of our church. The difference between my Sunday school teacher with her gentle voice and sweet smile and Dr. Lampe was stark, and shortly after my grandfather died I stopped attending church.

~~~

I have vivid memories of the exclusivity of our congregation and the lack of connection with the teachings of the bible, which certainly applied to all Christians and, I believed, to all mankind. As I grew older, I realized that there have been many religions and false prophets throughout history, who taught hatred of other races and/or religions and fought wars for their distorted beliefs. In my mind the current radical Islamic jihad, the terrorist Irish Republican Army and the crusades all have abused and distorted their religions.

There was a strong anti-Catholic element in our church that included my family. My grandparents and their grandparents had been in America for many years. My grandfather was a descendant of a Lord Lindsay from Scotland, who left under duress from the English and settled in Ulster (Northern Ireland). His descendants in turn emigrated to America in the early 1800s and the Williamsons settled in the Kentucky/Illinois area.

My grandmother's family were descendants of Bailey/Sullivan/Cross
~~~

families from Tennessee, Virginia, and Arkansas. The Williamson family were members of the "Old Lights" protestant religion but were ejected from their church when they voted. They joined the "New Lights" religion. These roots ran deep, including a strong bias against Catholics, and the family eventually ended up as Presbyterians.

Our congregation also had a strong bias against Jews and Blacks. I heard very little bias against Jews in my home, because we once shared (upstairs/downstairs) a house on Etzel Avenue with a Jewish family, the Fishmans, and the two families grew very close. I later discovered a family secret—my great-grandfather had married a Polish-Jew (Hissong) on the Williamson side of our family. I felt very good about it.

Prejudice against Blacks was barely noticeable because there were none in our neighborhood, school or church. The only blacks I knew growing up were Joe, our janitor at West Presbyterian church, and Buddy, a very big black man that I worked with at Carpenters Cookie Store, baking oatmeal cookies. Saint Louis was far more segregated than the South to the extent that we seldom saw any blacks in the neighborhoods. Thus, I inherited all of the myths and prejudices regarding the races and religions but grew up with and enjoyed my Catholic and Jewish friends in spite of my bias. I guess I maintained my bias against the tribes but ignored it for my friends who belonged to the tribe.

My deception worked with my friends but not their parents, who could see my bias through my thin facade.

As I grew older, I recognized that the Jewish families possessed a culture that was more wholesome and produced stronger families and values. My mother was not biased at all and never participated in the negative racial/religious dialog in our home. She was the only one in her family who had graduated from high school and the only one who shared the religious practices and values of my grandparents. Thus, I could learn positive values from observing the good, the gray and the bad values within my own family.

I had a hard time sorting out my own value system and defining good and bad values, because I loved all of my family and slowly realized that all of us had some good, bad and downright ugly values. I also learned that the public's interpretation of good, bad and downright ugly values changes with time, as public mores change.

Values and Ethics

I believe that the biggest change that has occurred is that my values were molded first by the family, second by my school teachers and church, and third by my peer group, with the peer group growing in influence as I aged. I believe modern children are influenced more by their peer groups and by television/ movies/ advertising/computer games than I was and that this tends to dilute the influence of the parents, teachers and the church.

I believe the modern family faces many more challenges in molding family values, as the media constantly reveals a plethora of information sans value judgments, and religious interpretations are discouraged by our agnostic society and by our own laws. We all have total freedom with little responsibility, and materialism is a dominant measure of value. We are denigrating the achievements of our past leaders, using our own distorted interpretations of what is currently considered politically correct. Even history has become a victim of political correctness.

My first encounter with ethics was to learn not to tell lies or exaggerate. Our family had a number of good storytellers, exaggerators, liars and a few good listeners. It took me a long time to know the differences, longer to understand the differences, and longer yet to recognize them. It was ages before I learned how to be a good listener. Remember that my childhood was back in the time when family conversation was the dominant entertainment and our dining room dinner table (and playing card table) was our theater. My grandfather was an excellent storyteller and told many fanciful tales about his life experiences. My grandmother provided clues to us young ones as to the degree of truth in the content but, they were usually over my head and I listened raptly to all of his wonderful stories.

My Uncles Dale and Don seldom talked or listened, but their conversations addressed the business at hand like "potatoes" meaning "Please pass the potatoes" or "Why in the hell did you bid so much?" during the family Pinochle games. When they related stories, they were not interesting to kids, and my mother would later reveal to me the degree of truth in the content. The uncles also were usually crabby.

My Aunt Bess told very funny and imaginative stories and generally was the life of the party. She loved people, particularly men and children, drank a lot and was very good entertainment.

My Aunt Gladys and my mother enjoyed the entire family, listened a lot and could be relied upon to help my grandmother with all of the family chores. They related stories about the children's positive and productive daily activity. They did not talk much. My great Aunts, Mamie and Eva, were frequent visitors and enjoyed being with our family. They loved us a lot and would greatly exaggerate any of our accomplishments. There were many other family members and friends who would come for dinner or to play cards or monopoly. My favorite spot was underneath the dining room table where I could listen to all the conversations until I fell asleep. As a result of all of my eavesdropping, I became a pretty imaginative storyteller myself.

Stories or Lies?

When I started at Dozier Grade School, I would walk home to 932 Maryville Avenue, cutting through neighborhood yards and occasionally picking flowers or using tools that had been left out. The neighbors complained and I was forbidden to cut through their yards. Well, I continued my shortcuts but went faster and was very stealthy. One of the yards had a barbwire fence, which I had successfully scaled many times. My increased speed, however, proved my undoing and I fell into the barbwire head first. By the time I worked my way out of it, I was a bloody mess with several shallow cuts in my head.

This required a very good explanation, so I ran the rest of the way home and told my grandmother and mother that I had been attacked by a lion on the streetcar tracks (a legitimate route home) and he had chased me home. While they cleaned me up I could see considerable skepticism, so I embellished the story some more, claiming to have seen several wild animals in our neighborhood in the middle of the night. They clearly did not believe me, so I proceeded further down the liars path offering several more supporting stories—each one less credible than the original lie. This went on for a day or so and it was clear that my mother and grandmother did not believe me! I finally felt very ashamed because I could see that they believed less and less of my story. So I finally confessed, stating that I was just trying to tell a really good story like my grandfather to hide the fact that I was still cutting through the neighbors' yards.

That was just one example of my exaggerations. I remember very vividly being trapped several times by my own lies. I can still feel the shame as the red flush came up my neck and around to my cheeks; and my eyes became big and round. My grandmother would get this disappointed look and be silent for a long time while my shame became agonizing. Before long the entire neighborhood and even strangers could easily tell when I was lying or even exaggerating just a little bit.

After trying very hard to become a credible liar by becoming a good actor, I finally figured out that it was much easier to tell the truth and get it over with! To this day it is very difficult for me to exaggerate even slightly to make a story more humorous, because I can feel my guilty look emerging. My grandmother cleverly installed a conscience into my soul; and, as a result, my stories are not as good or as funny as my peer grandfather group. Most are very dull as stories go.

Spiritual Progress

My spiritual development also was very slow. I was always a curious child (and adult). I remember receiving excellent credible answers to my many questions from my grandfather. He was a Bible reader and frequently brought Bible stories into play in his answers. My Sunday school teacher was also very good at answering my many questions on her Bible stories, and I probably made a pretty good straight man for her. I knew early on that everything spiritual depended on faith. I had to wrestle with that for more than 30 years before I could satisfy myself.

I always had a difficult time interpreting the many different Books of the Bible. Most of the stories required a strong faith to believe; some of the stories appeared to reward the sinners; and the massacre stories drowned everybody except Noah's family by flood, washed away the Egyptian cavalry with the Red Sea, and turned the people of an entire city into pillars of salt! When these stories were discussed by our sweet Sunday school teacher, they did not disturb me, and my grandfather could usually provide reasonable answers. However, when I was old enough to go to church services, our minister delivered the sermons, and his interpretations of these stories seemed inconsistent and overly harsh. Those killed by this vengeful God obviously faced hell or just ceased to exist. Surely there were innocent children, mothers, grandparents, and soldiers just following orders among the dead.

I remember developing a fear of death and could not imagine burning in hell or just ceasing to exist. I also remember having nightmares about dying. I knew that the Bible was important and my mother and grandparents believed that it was the key to heaven, but our minister made it appear to be very hard to get there. I believe that he was exploiting fear to get us to heaven and it was working on me.

I was 11 when my grandfather died; and, having lost my Bible scholar, I stopped going to Church. My nightmares stopped, and it was a long time before I returned to formal church services. I had started working as a paperboy when I was 11. By the time I was 12, World War II had drafted most of the young men away, and I started working as a delivery boy or soda jerk in our neighborhood stores. When my grandmother tried to take me to church, I would always have a Sunday morning job. It was the start of my work ethic but delayed the development of my value ethic. I was to learn and earn my value ethics the hard way—one mistake at a time in a less forgiving workplace than our family home.

Family Entertainment

All of our family read books a lot, and our living room almost always had one or more readers. My sister and I read a lot, usually in our bedroom. I loved Grimm's Fairy Tales, children's storybooks, Jack London's books, Captain Horatio Hornblower's books and books about philosophy, adventure, WWI war stories, and history. The books stimulated my imagination and curiosity and heavily influenced my pursuit of a life of adventure.

Our whole family would listen to the Sunday evening radio shows—Jack Benny, George and Gracie Allen, Fred Allen, The Great Gildersleeve. My grandmother was hooked on the daytime serial, One Man's Family, and we also tuned into THE SHADOW. My grandfather would listen to the St. Louis Cardinal and Browns baseball games. Radio was good because I developed visualization in my mind to supplement the story I was hearing. I doubt that we spent more than three hours per week listening to the radio. (We probably waste that much time now listening to TV commercials.) I also vividly remember Orson Wells' very real broadcast of our invasion by Martians. That scared all of us, and again I had nightmares. Looking back, I realize that I always enjoyed scary stories, books, radio shows and movies because they were exciting—but not real.

My childhood criteria for being real was belief by my family. I was really scared by the bogus Martian invasion (and the more gruesome Bible stories/sermons) because my family believed they were real. I was entertained and excited by scary books, stories and movies because, while they were much more credible, they were just stories. Thus my fear was a measure of my faith in my family. My nightmares must have meant that I did believe those Bible stories were real. So my family and Sunday school teachers must have instilled a good measure of childlike faith sufficient to produce nightmares. The seed of faith in God was planted.

My childhood was filled with fun enhanced by a spectacular imagination. Again, we had the advantage of possessing very few toys.

Tommy Whelan, Jackie Bell, Billy Bohan and I used kitchen knives and spoons to build roads and riverbeds on a 20-foot square plot of bare rich Missouri soil, which my grandfather reserved for our construction projects. We would use the garden hose to fill our "rivers" and build crude bridges over them.

We could play in our backyard when our bed sheets were hung on the clothesline to dry—we would sit in a bushel basket or big cardboard box fashioned as a boat and imagine that we were out to sea with our sails flapping in the wind.

Garbage can lids were our shields and broomsticks our swords, as we knights-of-old staged sword duels. As we grew older, we ventured out of our yards and fought other neighborhood tribes using big weedsticks pulled from the ground with a lethal tip of sod and roots as spears. Again, garbage can lids served as shields.

Later, as we acquired hand-me-down, screw-on-shoe roller skates, we added mobility to our games. We made scooters out of orange crates on a two-by-four with old roller-skate wheels and we had chariots. The neighborhood garbage pits in back of all houses on the alleys provided most of our material needs.

We were blessed with an absence of adult supervision, and we organized and played "sandlot" softball and touch football, with the two best players choosing up sides. Feelings were hurt when we were not chosen or when a better player showed up and we were replaced, but we always knew that our turn would come when we were bigger and better and could play as much as we wanted.

By now we were in school, and our circle of friends encompassed the whole neighborhood. We could go out to Forest Park and exercise our imaginations to the limit. We could also fish in Forest Park, which provided a rare opportunity for city kids. Naturally, we made our own fishing poles and used bread or worms as bait. Fishing in the big lagoon below the Art Museum was illegal, so we would hide in the bushes with the poles out over the water. One day, two of us were fishing and we heard someone coming around the bushes. It was a policeman and he had caught us red-handed! He looked at us carefully and asked, "What are you using for bait?" We replied, "Bread." He said in his Irish brogue, "It's a pity because using bread is a Federal offense! I'll have to take you both in." He was old and rather fat, and he started coming toward us, going around the bush until he was very close to the water and us. He made it obvious that he was having a difficult time and we took the cue and ran for our lives, never looking back. It was years before I figured out that he was having fun playing his game. I wonder if he took our fishing poles home?

The policemen who walked our neighborhood beat were kind of authoritative grandfathers. When they were called because people were tired of us scavenging their garbage pits and borrowing their garbage can lids, they approached noisily so that we could run ... and quietly, if they wanted to give us a stern lecture.

In the winter we took our sleds to Art Hill in Forest Park and used our screw-on ice skates in the lagoon. Usually there was a big bonfire where you could warm up. There were always slushy days when you could stay home and read books. In the autumn we had plenty of sycamore, maple and oak leaves to play in and the colors were wonderful. Springtime was always welcome, and we were outdoors most of the time when we were not in school.

Summer was best of all, but I remember the very hot nights when we would go out on the porch roof, hoping for a breeze to cool our sweaty bodies. My mother, sister and I shared a bedroom and all of us would lie awake in the heat. I guess we did not have a fan in our room.

Regardless of the heat, we all loved summer because there was no school and a lot of play, and usually a summer vacation in Silver Lake, Wisconsin, or Bagnell Dam in the Missouri Ozarks. All of our expanded family group would rent a cabin— there must have been 20 of us. Most would sleep in the attic, and the grandparents and great aunts would sleep on the first floor. We could fish, swim, play cards or Monopoly, row boats or canoes and have wonderful cook-out dinners. Cousin Bee taught me to swim in Silver Lake. Aunt Bess would row the rowboat while I swam alongside, until I was able to swim all the way across the lake. Best of all, I could go exploring in the woods, hiking all the trails, always curious to look for critters and see what was around the next bend. I can well remember being bit twice by non-poisonous snakes that I grabbed by the tail. (I never learned on the first mistake!) The dinner table was our best time, with good conversation, jocularity and very good food. Our family always had fun, and even Dale and Don would join in the vacation merriment.

Aunt Bess and Mother, 1939

Most of my friends loved airplanes. We read everything we could find on aviation, and all of us built model airplanes in our homes.

On Saturday mornings the big department stores in downtown Saint Louis (Famous & Barr, Boyds, Stix Baer and Fuller) would sponsor model-building classes, and 20 to 30 kids would work side by side, cutting out the balsa-wood ribs and stringers; gluing the wing, fuselage, rudder and elevator assemblies together; and then covering the assemblies with paper. We would put clear dope on the paper, and when it dried we would assemble the parts to form our aircraft, install the landing gear and connect the propeller with a strong rubber band strung from the prop through the fuselage to a strong wooden cross section in the tail. We would then balance the aircraft so that the center of gravity was slightly forward of the thickest part of the wing (center of lift). We would throw it gently to see how it would glide and trim the elevator to achieve a low dive angle. Then we would put in a little rudder trim to offset the torque caused by the rotating propeller. Finally we would take it out to a field, wind up the prop, give it a toss and watch it fly. Most of our models flew well after a few adjustments.

I remember always having five or six models hanging from the ceiling in our bedroom. I made models of all of the World War I German, French and British fighter aircraft as well as models of the current racers and private planes. Most of them eventually crashed, and after they had been repaired many times we would put a little dope on them, light them up and let them fly away, burning to a glorious end.

Changing Times

My childhood probably ended with the death of my grandfather, Jacob Lindsay Williamson, in the summer of 1941 when I was 11 and my sister Doris was 15. Our strong male leader and sole wage earner was dead, and many changes were soon to occur. Only now do I realize that this began a critical transition in my young life. Our country was at war, and all families were entering a critical transition period that would change our family structure, mores and our interface with the rest of the world. Change would dominate our country for the next five years.

Me and My Sister, 1939

~ CHAPTER 2 ~

Growing Up

My Grandfather, 1939

I enjoyed every aspect of my childhood, and I can mark the end of it with the death of my grandfather, Jacob Lindsay Williamson, on 17 June 1941, when I was 11 years old. We were living in a big house at 5595 Chamberlain on the corner of Chamberlain and Clara, and we had just celebrated my grandparent's 50th wedding anniversary on 21 March. They had the only successful marriage in the family, and my sister and I wanted to be as happy as they were. After grandfather died, we took in a border to help pay the rent. I remember that we could have bought the house for $5,000, but that was never a possibility.

We had always rented for about $50 per month. My aunt Gladys, who was working in the cafeteria at Monsanto Chemical Company, assumed the burden of supporting my grandmother, my mom, my sister and me. My grandmother received a social security check of about $30 per month. The money would not stretch far enough, and we had to move to a smaller house at 5744 Bartmer, just three blocks away, which we rented for $35 per month. Dale continued to live with us but did not pay his share of the household expenses. He did, however, prepare a delicious beef stew for the family dinner once a week. Sometime before the end of World War II, both Dale and Gladys bought a car, and that was a first for our family. I can remember Gladys taking my mother, sister and me on our yearly family vacations, but that was the only time that I can remember being in a family car. The cars were used by Gladys and Dale to go to and from work.

I guess we were poor, but I never felt poor because we had a lot of family in our home, also a lot of family elsewhere in our city and in the nearby country—in Missouri and across the river in Illinois. I also can remember that jobless drifters would occasionally ring our doorbell and offer to work for food. My grandmother would serve them a plate of food on the back porch, and sometimes my grandmother and I would talk with them. That must have given me a feeling of wealth. My sister, Doris, did feel poor, but I think that was because she had been deserted.

The Novice

I started working part time as a paperboy when I was eleven, after grandfather died. I worked for a big, mean, dapper man who had several paper routes and the selling-rights to several good street corners along Delmar, a busy business street in the West End of St. Louis. Several boys from about age eleven up to sixteen worked for him part time. As one of the youngest, I occasionally got to deliver papers on Sunday, and I occasionally was allowed to street-sell the corner of Belt and Delmar. We would hustle the corner near the streetcar stop, carrying the paper bag and shouting, "Read all about it, read all about it! Police raid West End bootleg house!" or whatever was the current headline. I cannot remember our pay arrangement, but I do remember that it would be a couple of nickels and some pennies.

Slugger

St Louis had three newspapers in those days—the Globe Democrat (morning), Star Times (afternoon), and the Post Dispatch (evening). Often there would be special editions for big stories. The three papers were very competitive, and overlap began to occur across the whole day.

I also began singing in the boy's soprano choir in the Episcopal Church on Goodfellow near Cabanne. They paid us 50 cents per performance and two dollars if we sang a solo. (I never soloed.) I remember that I would make more money singing on Sunday than working as a paper boy, and the choir master was a lot nicer than our dapper paper-boy boss who threatened us regularly. Like all good things, the choir job disappeared tragically at age 13, when my voice changed while stretching for a high note singing "Panis Angelicus." It was the beginning of my education in economics, personnel management and aging gracefully.

The Odd Jobs

I think I worked for every small business in our neighborhood before I was 15 years old. I was a delivery boy and soda jerk for Hoskins Drug store and Stones Drug store; baker's helper and soda jerk for Carpenters Ice Cream and Oatmeal Cookie store; delivery boy and helper at Brennan's Grocery store; soda jerk at three different Green Leaf Ice Cream stores; sold soda pop in the aisles at the municipal opera in Forest Park; and a theatre usher at the Plymouth, Esquire and Pageant Theatres.

I started out working for five cents per hour at Carpenters and worked my way up to 25 cents per hour in the menial store jobs. However, I then broke into the big time in theatre jobs, which paid 50 cents per hour! I probably worked about 12 to 15 hours per week from age eleven to fifteen and then made head usher at the Pageant theatre at 75 cents per hour until I was a senior in High School. I worked about 20-28 hours per week at the Pageant. Because I refused to join the union and pay union dues after they unionized the theatre ushers, I was fired from the Pageant job. It was a very humiliating experience and I never forgot it. That was the end of my career in show business, but I had seen most of the movies made between 1944 and 1947 four to six times. Admission costs were a dime, when I started and 25 cents when I was fired. The admission included a feature movie, a class-B movie, a newsreel and a cartoon.

I do remember that some movies glamorized criminals; and Jimmy Cagney, George Raft and Humphrey Bogart played many criminal roles. I also remember that the movies, radio and newspapers were all on our side all during World War II. Of course we had Hitler and Tojo, who were ideal villains, as enemies during World War II. It was wonderfully patriotic, and most young men wanted to enlist and fight the Nazis and Japanese. We all bought war stamps at 25 cents each, which became $25 War Bonds when we accumulated $18.75 (75 War Stamps in a book).

That was a far cry from today, where the media, academia and Hollywood tend to be biased against their own President and their military. I seldom attend movies now. They lack a storyline and work foul language into the scripts like a small child trying to get attention. They also show excessive, unnecessary violence and glamorize sexual promiscuity without revealing the real world consequences.

I was always very curious and had a vivid imagination. I also craved exploration and adventure. I was very impressed with war movies and warplanes, making models of all of the World War I fighters before my grandfather died. My Uncle Dale implied he had been a pilot in France during WWI, although I later found out he was in aircraft maintenance. My mother had told me that my father had been a Captain in the Royal Irish Rifles and had been wounded with mustard gas during WWI. (Much later I found that he had been an enlisted rifleman in the Ulster Division and had been buried in a trench by German artillery at the battle of Somme but was dug out by his comrades). Anyway, I was very anxious to go to war, fight for my country and return home as a hero with many medals and battles. My imagination ran wild, and I daydreamed about future combat exploits in school and at nighttime before I fell asleep.

Movies and adventure books fueled my imagination. I remember being impressed by the philosophy of Socrates facing his death (as related by Plato) compared to current religious beliefs, and the contrast between the carefree Trojans and the well-disciplined Spartans in Homer's Iliad war story. My destiny was slowly beginning to take shape, and I knew that I would have to grow tougher and improve my physical conditioning to become a ... ***Hero.***

Really Odd Job: The Gandy Dancer

When I was 15 and summer vacation finally arrived, I applied for, and was hired by, the Wabash Railroad as a railroad laborer (gandy dancer). Seven other boys and I traveled by train up to Muskegon, Michigan, and then took a large ferry boat across Lake Michigan up to the middle of the upper peninsula of Michigan to work on the 30 miles of railroad between Manistique on Lake Michigan and Manistee on Lake Superior.

We lived in a one-car garage in Mrs. McManus's Boarding House in Steuben, Michigan, midway between Manistique and Manistee. There was one dirt-and-gravel road and one rail line that ran through Steuben, which boasted a population of 13. There was no running water, so we washed in the fast cold creek in back of the house or pumped up a bucket of water from the well and heated it on our wood-burning, pot-bellied stove. We used a two-hole outhouse and I hated it.

I was the youngest, but I was already over six feet tall and could work as hard as the others. Our foreman was Mr. Burley. He had two other local full-time laborers augmented in the summer months with high school boys from St. Louis. Our crew would replace old railroad ties with new ones and occasionally have to cut and slip the tracks to increase the spacing between the rails that expanded in the hot summer sun. It was hard work and we did it nine hours per day with an hour for lunch, five days a week and a half-day on Saturday. I believe we made 75 cents per hour but we had free room and board, so I saved a lot of money.

Poor Mrs. McManus had eight of the hungriest teenagers that ever lived. I would eat 8 to 10 eggs and several pancakes for breakfast. She packed us a lunch for the road and I would eat 6 to 8 sandwiches. Her dinners were spectacular! We ate a lot of venison, trout, game birds and occasionally bear, always with a lot of mashed or boiled potatoes and vegetables. There were four lumberjacks and a forest ranger, who roomed in her house, and they were all hunters and fishermen. They taught us a good bit about living and hiking in the woods.

I was very ignorant about the facts of life and was leery about using the common outhouse for fear of contracting a disease from the toilet seat. I would sneak out into the woods to go privately and use catalog paper. One day I forgot the paper and used some leaves off of a bush. You guessed it—poison ivy or oak! I could not work for three days. My foreman said he had a sure cure. Anxiously, I encouraged it. He poured alcohol on my inflamed rear end, and I was certain that I would never be able to use any of that apparatus again. I suffered for another day and went back to work the next day. When the foreman inquired about my status, I hastily agreed that it was truly a miracle cure. I did not want a repeat treatment. Don't try it!

I loved the Upper Peninsula of Michigan. We could walk to a half-dozen lakes that had only a path through the woods with no roads or houses, and we could occasionally see the northern lights at night. We swam and fished in the lakes and the creek and hiked to our hearts content. We would see a lot of deer and occasionally bears, which we avoided. We could take the freight train into Manistique or Manistee on the weekend and go swimming in Lake Michigan or the very cold Lake Superior, or go to the quarry in Manistique and dive off the cliff into the clear deep water in the quarry.

I remember being in Manistique when we heard that the war was over in Europe. We celebrated by drinking some beer and spending the night in Manistique. We had no place to stay, so we asked the local policeman for ideas. He said, "Spit on the sidewalk and we will put you up in our jail." We did, and he did! It was a wonderful day when World War II was finally over and everyone was celebrating. When summer was over, we happily returned home to our families. I felt more grown up—that I had kind of proven myself, Raw nature and hard work were truly wonderful for a city boy. I really felt the call of the wild.

Very Odd Job: The Seal Blubberer

One of my fellow teenager railroad workers in Upper Michigan told me about his brother who worked for the Fouke Fur Company in St. Louis and who went to Alaska every summer on a sealing expedition. Holy mackerel! Fulfillment of my Jack London fantasies—an ***expedition in Alaska!***

As soon as I returned to St. Louis, I applied for the next summer's expedition. Willing applicants were given a very rugged tryout every December. It consisted of working the blubber off of as many sealskins as possible in four three-hour days, without damaging the hide. The higher your score, the better your chances. The sealskins had been preserved in salt and were very difficult to work. Our hands would blister and our blisters would blister! Many did not complete the try out. The Company knew I was only 15 when I took the try out but never suspected that I would make the crew. I was one of the five high school boys and five college boys who made the team. Just before we were scheduled to leave, they rediscovered that I was only 16 and they had to check to see if they could take me. Thank heavens they could. So, in late May 1946, I left St. Louis on a real expedition to the Far North. It was the first of three summer sealing expeditions that I enjoyed with the Fouke Fur Company,

We traveled by train overnight in a sleeper car up to St. Paul, Minnesota, and then transfer to a Northern Pacific sleeper car for a wonderful three-night journey across the scenic northern border of the United States to Seattle Washington. We enjoyed this luxury and ate three meals every day in a dining car with white tablecloths and silver utensils, waited on by courteous black stewards. All of us wore a suit with tie and were firmly instructed to act like gentlemen to preserve the reputation of the Fouke Fur Company. (I had to invest in a suit, shirt and tie before I departed.) We were given detailed lists of the items and work clothing to be packed, and I believe we were given a reasonable clothing allowance.

We stayed in the Olympic Hotel in Seattle for one or two nights before boarding our ship in Green Lake. The suits came off and the luxuries ended after that wonderful stay in Seattle. We traveled to and from St. Paul Island in the Pribilofs, located above the tree line in the middle of the Bering Sea, on one of three ships operated by the U.S. Fish and Wildlife Service. On my first and third expedition we used the Penguin, a 115-foot twin-engine, two-masted small cargo ship with a crew of about eight and our Fouke Fur crew of about 20 blubberers and three supervisors. On my second expedition we used the Black Douglas, which was converted from a 105-foot, three-masted sailing yacht to a twin-engine, two-masted small cargo ship with a new superstructure above the deck. A patriotic American had sold his yacht, the Black Douglas, to the U.S. Government for $1 and they had converted it to a small cargo ship.

Going to sea on these small ships was a real adventure. We would start our Odyssey in Green Lake, traversing the canals through Seattle and going through the locks into Puget Sound. We would then go into the open sea for a short time before entering the inland waterway that would take us up the Canadian and Alaskan coastline to Sitka, Alaska. Most of the inland waterway was sheltered from the Pacific Ocean by offshore islands.

I will never forget my first introduction to the open Pacific Ocean aboard the Penguin on my first expedition. John Carmichael and I still had our suits

Seal blubberers on the *Penguin*

on and stayed up in the bow after we stowed our gear in the foc'sle. We did not want to miss a thing and, despite the urging of others to change our clothes, we were to eager to experience all aspects of our adventure. As we left Puget Sound, we began pitching in the ocean swells. Soon we were firmly holding the rail and waves were breaking over the bow drenching us in our suits. While this was great fun, we could not get our suits cleaned and pressed until we returned to Seattle eleven weeks later. My suit never looked the same. Later, when the ship was rolling, John and I went out on deck at night and climbed up the rope ladders on the aft mast. One of us took the port ladder and one the starboard ladder. We climbed halfway up and took a firm grip. As the ship rolled from side to side, the slack in the ladder would snap from side to side, pressing one of us against the ladder while the other was hanging on for dear life to keep from being thrown overboard. Then she would roll the other way and we would snap in the opposite direction. Back and forth we went laughing uncontrollably and thoroughly enjoying our very dangerous and stupid game.

We spent a lot of time on deck while traversing the inland waterway. There were a lot of Orcas (killer whales), whales, and porpoises to watch, heavily timbered mountains on both sides of us and many different kinds of birds from eagles to sea gulls. It was a luxurious panorama of raw wilderness. I fell in love with Mother Nature in Alaska. (Fifty years later my wife and I took the same voyage on a Princess Cruise ship. We sighted far fewer whales, but Mother Nature had preserved all of the rest.)

Once we left the protection of the inland waterway, the Captain would decide which route to take from Sitka to Dutch Harbor, which was midway out the chain of the Aleutian Islands.

The northern Gulf of Alaska poses a challenge to small ships. If the weather and sea state were normal, we would go directly across the Gulf to Dutch Harbor. If not, we would take an extra day and go between the Aleutian chain and Kodiak Island. This was even better wilderness because we could occasionally see the giant Kodiak bears on the shore. On my three round trips (six voyages) I only remember taking the Kodiak long route twice.

I will never forget the trips across the Gulf of Alaska. I have never seen larger waves. When we were in normal weather, we would roll or pitch constantly and all of the blubbering crew— except John Carmichael and me— would be miserably seasick and stay in their bunks. The cook could not prepare hot food, so we ate sandwiches and held on to our sandwich and our drink.

Everything had to be tied down or stowed. When we were in bad weather with 30« to 50-foot waves and pitching, the screws would come out of the water and the entire ship would shudder and vibrate as the screws speeded up without the resistance of the water. When we were rolling, we would hit 25 to 35 degrees, and sometimes it would seem that we were on the brink of rolling all the way over. When we were down in the trough of a wave, the advancing peak of the next wave looked higher than our mast, and it was miraculous that we would rapidly rise to the top of the wave and then come racing down into the trough again. We would cruise through the water, pitching and rolling and bobbing up and down like a cork! It was far more violent than the roller coaster in Forest Park, with a lot less certainty that you would stay afloat. We had heavy lines out on the deck that we would grasp and walk along whenever we had to go out

The *Penguin* and Me

there. The decks were awash much of the time and the salt spray was driven by the wind across the deck. John and I loved the action but were very careful whenever we went out on deck. In the Gulf of Alaska much of the deck was beyond our definition of fun, but it certainly qualified as adventure. It felt very good when it became more calm; and, we were always glad to enter the port of Dutch Harbor or Unalaska to refuel and spend the night.

The last leg of our voyage was the 250-mile voyage through the fog-shrouded Bering Sea to our final destination, St. Paul Island, Alaska, in the Pribilof Islands. The wind was almost always more than 30 knots, and the fog would race along 100 to 200 feet above the Bering Sea, hugging the ground when the wind velocity slowed down. The Bering Sea was very rough but not as rough as the Gulf of Alaska.

The Russian navigator, Gerasim Pribilof, discovered the islands back in 1786 and claimed them for his country. An old Aleut had told him about the uninhabited seal islands north of the Aleut village that is now called Dutch Harbor. He sailed north into the fog, and after several days they could hear the seals bleating and barking before they could see the island. He had found the southernmost island of St. George. The fog was so thick that he did not find St. Paul (the largest island), just 8 miles north, until his second voyage. He enslaved the Aleuts to hunt and skin the profitable fur seals and transported many of the Aleuts north to work on these barren volcanic islands in the middle of the Bering Sea and above the tree line. Many of the natives died in this workplace.

We always had a research team of biologists from Washington University in Seattle who would accompany us on our voyages to and from the Pribilofs. They would tag some of the seals, estimate the size of the herd and monitor the health of the seals. On my last expedition in 1948, a Walt Disney film team also accompanied us on our voyage to film a documentary on the fur seal. I never had the opportunity to view the results.

During the 1940s about four million fur seals migrated to the Pribilof Islands every spring after the ice pack melted. They came to breed and to bear their seal pups that were conceived the previous year. The female has two wombs and can carry last year's pup in one and become pregnant in the other. The pups must be born on land and cannot swim until taught by their mother. There are several rookeries (breeding places) on the islands with most of them on St. Paul Island.

The larger bulls are the strongest swimmers and are the first to arrive. They select their location on the rocky beach and defend their turf from all intruders. When the females begin to arrive, the bulls attract them into their harems, often fighting over a female and sometimes killing the female during the fight. By the time we arrived on St. Paul, the rookeries would be packed with many harems and millions of fur seals barking, bearing pups, breeding and sometimes fighting. Perhaps 50 or so Stellar Sea Lions and an occasional Walrus, both much larger than fur seals, also were in the rookeries.

The Aleuts would then begin the daily kills before sunrise. The seal herd was controlled by the U.S. Department of Commerce, Fish and Wildlife Service, and they only allowed three-year-old male fur seals to be killed. They were still bachelors and would hang out together like young men at the corner bar, hoping to pick up a stray female from an older bull's harem. Since all of the fur seals were separated in age by one year, it was easy to differentiate two-, three- and four-year-old males and the bachelors, grouped together apart from the harems and their very alert bulls. The Aleuts would herd the bachelors over to a killing area to kill the three-year olds and strip the hides off of their bodies. They would then load the hides into a truck and deliver them to the Fouke Fur Company blubbering shack. The carcasses would be hauled to another building and ground up to make fertilizer after the fins and livers (used as food) were removed. We would blubber and salt-down 55,000 to 70,000 fur seal hides during our 10-week stay on St. Paul Island.

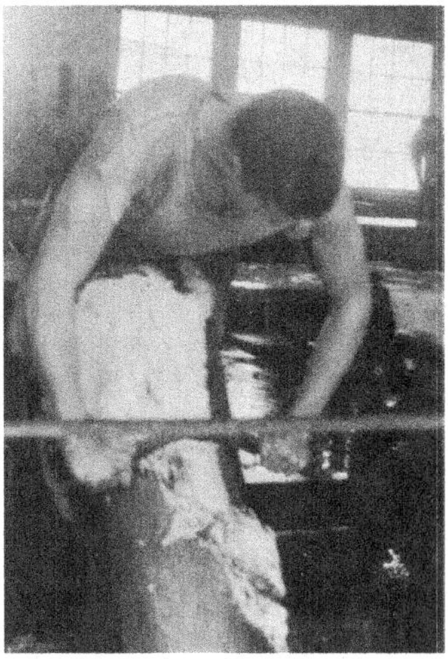
Bob: Blubbering Seal

The blubbering work was very hard and most days we would blubber two to three hours per day and then salt down the hides in the salt house for three to four hours. The natives would do one rookery every day. There were two very large rookeries on St. Paul. On those days it would take us four to five hours to blubber the kill, and we would only be able to salt-down half of the hides. They seldom made us blubber more than four hours, because we would become punchy bobbing up and down, using our upper body weight to stretch our long blubbering stroke down the length of the sealskin. It was very hard work but we were well paid. The young student blubberers would make about $200 a month with all room, board and travel paid for the 14 weeks we were away from home. At Christmas time we would receive a bonus of seven cents per skin or about $250 to $300. I would save almost all of it; and I could handle all of my expenses for the next year and use my income from ushering for spending money.

There were about 200 to 300 Aleuts on the islands and the U.S. Government ran the village as a commune with a U.S. Commissar, a medical doctor, and Father Baronoff, a Russian Orthodox Priest. The native children attended high school in Sitka, commuting via the three Fish and Wildlife ships. The Fouke Fur Company paid for all of the native labor to the U.S. Government, and the commissar distributed the money to the native laborers. It was a socialistic system run by our government. I believe that it changed after the cold war started.

Fouke Fur Company:
Blubbering Shack, Salt Bins, Storage Building Bunkhouse (in front)

St. Paul was a volcanic island with a dead crater not far from the village. The island was rocky with very little soil, mostly surrounded by steep cliffs that rose out of the cold Bering Sea. There were occasional rocky beaches, and the larger ones were the fur seal mating grounds. Inland there was thin tall grass and many kinds of wild flowers that flourished while we were there. Some time ago the government had relocated a reindeer herd to the island, hoping to provide a food source for the Aleuts. We would occasionally come across their antlers and skeletal remains when we hiked around the island. I never saw a live reindeer on the island, and I suspect that they all fell into crevices on the rocky terrain or starved during the winter months. There were blue foxes on the two larger islands and large rookeries of cormorants, sea murres, sea parrots and gulls.

The steep cliffs surrounding St. Paul were home to thousands of sea murres. The murres would nest and lay their large eggs on the flat rocks of the cliff. When the murres were away, the eggs (blown by the wind) would roll in

a perfect circle and remain in the same place. The murres resembled Penguins, with a black back and white breast. They had a large body and relatively small wings. When they left their nests, they would dive straight down, flapping their wings rapidly in a power dive to obtain sufficient speed so that their tiny wings would support their weight. As they obtained flying speed near the bottom of the cliff, they would pull out and barely clear the incoming waves breaking on the beach. Some of them would be offloading weight via bowel movements as they dove. There were always a few who would not make it and crash headlong into a wave. They would shake their heads and then perform a long water takeoff, flapping furiously and even running with their feet to lift that heavy body into the sky. They were fun to watch but aerodynamically challenged. They would sleep with their white breasts facing the dark cliffs and their black backs facing the sea so they blended into the color of the cliffs.

When the Disney film crew was aboard, the Captain of the Penguin anchored close to the cliffs offshore a few hours before sunrise and told his passengers if they wanted a treat to come out on deck with their cameras before sunrise. (We only had a couple hours of darkness during the summer months.) When the sun peeked over the horizon, the Captain blew his foghorn. Hundreds of sea murres awoke, turned around, and dove off of the cliffs. It was an awesome sight—the cliffs changed from a dark gray to a speckled white, as the murres turned around, revealing their white breasts and then dove off the cliff. Our "oohs" and "ahs" had to have pleased our Captain until the struggling, pooping murres began crashing into the side of the ship, or even worse, made it over the ship leaving a trail of poop as they crossed. Those of us who laughed the loudest were privileged to swab up the mess. It was worth it.

The younger blubberers would take long hikes all over St. Paul Island. We were north of the tree line and it always seemed odd to be in a wilderness without trees. We would find many kinds of wildflowers, walrus tusks, seal teeth and reindeer antlers and best of all "black diamonds" in the inactive volcano crater. Black diamonds were small, glossy black crystals that probably

Mount Bogalof, where I found the "Black Diamonds"

were a form of carbon that had crystallized under very high pressure and temperature when the volcano was active. I still have mine and still consider them part of my boyhood treasures.

John Carmichael and I even tried to swim in the frigid Bering Sea. The daytime air temperature was in the 60s and 70s, and we kept our jeans and sweatshirts on. We did immerse ourselves, swim a few strokes and head for the rocky beach. It was very cold, and the wind kept us cold as we ran back to our bunkhouse.

All of the 20 blubberers roomed in the bunkhouse in two-man, upper-lower bunks. We slept eight to a room and all of the rooms opened up into a big common room. We had a well-stocked library with very good educational, historical, mystery and adventure books. Reading was the dominant pastime. We had an occasional nickel-dime poker game. I loved to play and generally won or lost very little. I learned to bet from the older blubberers. We also had good and meaningful conversations among ourselves. Liquor was forbidden on the island and the company rigorously enforced this rule. The Aleuts occasionally drank raisin-jack.

I will never forget the Novak brothers. Cappy Novak was the oldest, maybe 50, and Frank was a few years younger. They could produce more perfect skins (no flays from the long, curved, two-handed blubbering knife) than any of us and took great pride in their work. They watched over us without being obvious and gave us a lot of good advice. The advice would generally be about some of the mistakes they had made in life and would generally come at a time when we were considering making mistakes ourselves. Working relationships, crew discipline, temper problems and some wonderful stories about their earlier lives really taught me a lot. On my third expedition I learned that they had been in a Chicago gang (maybe Bugsy Siegel's) and Frank had served time in an Illinois State Prison. They were straight arrows during the time that I knew them. I had not been exposed to male guidance since my grandfather, and the older blubberers were intelligent and very sensible. We had good discussions about philosophy and many different topics that were better than I experienced later in college. I probably listened better because we worked very hard, went to bed early and slept like rocks.

One of our college men was an ex-GI named Ted Eggerding, and he worked to earn and save money to last him through the college year. Ted played the piano and would lead us in song sessions. He picked four of the high school boys and formed us into a Barber Shop "Quintet." Ray Ogle sang lead, King Kong Craig sang bass, Squeaky Bayer sang soprano and Ted and I sang tenor. We memorized several old goodies and practiced our harmony sailing up the inland waterway. We made our debut in our first port, Ketchikan, Alaska. We docked in the late afternoon for an overnight and the five of us walked up the main street until we found a bar with a piano—less than two blocks from our ship. We went in, Ted sat down at the piano and we started singing. We were an instant hit in that entertainment-starved town. We sang well into the night, and the customers supplied us with free beer and sandwiches. The bar became packed and soon everyone was singing. Almost all of the population was men and there were no cruise ships or entertainment on the waterfront in those days.

We repeated our performance a few days later in Juneau, a larger port. Sometime during the evening a nice older lady (probably about 40) asked us to come across the bay to sing in her restaurant in Douglas, Alaska. We joined a large group and rode over to Douglas in the back of a pick-up truck. Again, we were supplied with beer and sandwiches and noticed that there were many more ladies in Douglas and it was a bar and not a restaurant. Before an hour passed we figured it out—we were singing in a brothel! Well, we were not the first to do so and undoubtedly will not be the last. We sang well into the night and had a wonderful time, and the fishermen who drove us there returned us to our ship before dawn. We repeated our performance in Sitka and Unalaska but did not come across any more brothels. It was a fun experience but you will not find it in my resume. Odd that you find it in my life-story.

On my second and third expeditions I asked and received permission from our boss to trade in my train ticket from Seattle to St. Louis for a ticket via another route. I paid the extra fare. We had always gone to Seattle via St. Paul, Minnesota, and then took the Northern Pacific Railroad to Seattle. At the end of my second expedition, I came home via Union Pacific—right across the center of the United States. My sister, Doris, had married Phillip Jackson while I was in the Pribilofs, so on my third expedition I went from Seattle to Medford, Oregon, to meet Phil's family and then down the coast to Los Angeles to spend time with Doris and Phil on their honeymoon. We spent one night on Laguna Beach and I swam in the spectacular crashing surf most of the night. It was a real thrill. I immediately liked my new brother, Phillip, and he tolerated me very well. He was a kind gentleman and it had never dawned on me that little brothers should not invite themselves to their sister's honeymoon. We shared many laughs on that later in life.

Phil and Me, Laguna Beach, 1947

I went home after a few days via the Southern Pacific route from Los Angeles to St. Louis. I loved riding on the sleeper cars and enjoyed the scenery, far better than the scenery from an automobile. By the time I was eighteen I had journeyed the northern, central and southern routes across the United States, the Pacific coast route from Seattle to Los Angeles, and the St. Louis to St. Paul route.

When I started traveling at age 15, I had a list of my father's brothers' names and would go through the phone book everywhere the train would stop and call every one of the name matches that I found. Miraculously I found his brother, Jim, in St. Paul, Minnesota, and he brought his sons down to meet me at the train station on my return trip. We corresponded for a while and he told me that he had not seen my father since he saw him with my mother many years earlier. He believed that he was dead. He also put me in contact with his brother, Bill, in Philadelphia and Bill told the same story. Many years later I learned that they had been in contact with my father and he had visited them during that period.

School "Daze"

I went to Dozier grade school for eight years and am still in contact with four good friends whom I played with—Jack Dalton, Dave Snyder, Bob Tully, and Ned Inglish. We had very good teachers and I can remember all of them. I also remember that I did not do the homework and daydreamed a lot while gazing out of the window. I excelled on the playground at recess and during our sandlot softball and corkball games. My mother must have been very frustrated because I was just an average student. Indeed, you could say that I was happy just being in school with lots of classmates and just getting by. Mother was upset because a college girl in our neighborhood would give me, and others, an IQ test periodically as part of her graduate project. She told my mother that I was a near genius with an IQ of 135. Well, mother told me and I figured out that I could always get by with minimum work.

Mother's expectations were not high until my IQ was measured. I was very slow to learn to talk and could not be understood until I was four. (My sister claimed I did not need to talk with all of the doting females in our house. She claimed I just pointed and said "ugh" and one of my loving family females would rush to bring the item to me.) I was scholastically lazy and preferred to look out the window. The exception was arithmetic, science and history, which I always enjoyed but seldom studied.

I enjoyed my classmates probably more than they enjoyed me. I could always sense that the kids (who had both parents) and their parents tended to discourage contact with me. I think I was the only boy who worked part-time in Dozier Grade School, and I was very visible working in all of the neighborhood stores. They may have figured that I would soon be among the many single-parent kids who dropped out of school as soon as they could. I do not remember feeling bad about that, but I usually felt awkward when my scout troop held Dad's day and some dads would show up to review their son's scout projects. Since there were quite a few boys without fathers during the depression, we were not unique, at least in rental neighborhoods.

I went to Ben Blewett High School and knew that I would have to do better scholastically. I selected a rigorous pre-engineering curriculum that included four years of math, four years of science (with chemistry and physics), four years of (ugh!) English including speech, two years of German, three years of history, mechanical drawing, and architectural drawing. The results were the same. I graduated right in the middle of my class. But I had an excuse: Not only did I work—I discovered girls!

It started in my freshman year. I went to parties where we played "Spin the Bottle" to select which boy would take which girl into the coat closet and kiss her right on the lips. I kissed Marilyn Smith, Carolyn Branch, Alta Morehead, Mary Perrine and others during these parties and enjoyed doing so. In our sophomore year we started having hayrides out in the country and occasionally we'd "neck" in the fresh-smelling hay. I started dating in my junior year and fell in and out of love with several very nice girls who became very good friends.

Sue Nicholas was my first puppy love and I had to compete with my best friend, Jack Dalton. He won for a while but my persistence eventually resulted in many dates. We would mostly walk around the neighborhood at night, have a soda at the Parkmore drive-in, and then I would walk her home. After several of these dates she allowed me to kiss her goodnight on her doorstep. It was my first non-party, non-game kiss. I remember running home across the neighborhood yards hurdling the hedges. I also dated Merrill Keogh, Nancy Lichty, and Barbara Siegel, all of whom were also very good friends even after high school.

Jack Dalton and I were best friends and we confided everything to each other. We both lettered in track and were ushers at the Pageant theatre. Jack was one year ahead of me, studied hard and made good grades. As a result of my friendship with Jack, I became more ambitious and began to worry about how I could ever get into, or pay for, college. I could see Jack winning a scholarship, but my grades would never make it. We both were determined to make $10,000 a year someday. (That is the equivalent of more than $100,000 a year in 2000.) I started studying harder, but it was going to be a tough sale to enter a college.

Earlier, I had discussed quitting school with my homeroom advisor, Miss Grace Mulholland. I was impressed with my earnings with the Fouke Fur Company and knew I could make even more money until I was old enough to be drafted, become a war hero and then find even better jobs. Miss Mulholland talked to me for a long time and we figured out that I wanted to be a pilot and an engineer. She then described the educational requirements and pointed out that I was already taking all of the right classes to pursue my objectives but needed to make better grades. She knew I could do better and conclusively convinced me that quitting school at that time was totally contrary to my objectives. I loved and admired that lady who introduced me to seeking objectives, determining the requirements and making a plan to achieve them. I wish that all young people could have teachers as smart and caring as Miss Mulholland. Several years later I went to see her before I went off to pilot training and thanked her. She knew that I was spending too much time working after school, daydreaming and chasing girls. I knew I needed to focus my energy on school, but the call of the wild continued to prevail.

I was 76 inches tall before I was 15 and had finally grown through my awkward stage. I was also very competitive and could conjure up my adrenaline when I needed to excel. When I really needed to perform to stay ahead, I would feel the rush of adrenaline, my energy would surge and I would win. Unfortunately, prior to attaining the age of 15, I was awkward (and tactless). I had two broken arms, one broken collarbone, one broken leg, several nose injuries without resetting the bone, a slight concussion and numerous sprained joints. I had developed my competitiveness, adrenaline pump and aggressiveness well before I developed grace.

I joined the track team and the basketball team when I was 15. I was the school's most competitive junior athlete in the 120-yard low hurdles, the high jump, the shot put, the discus throw and the broad jump. In the annual City meet I took first place in the low hurdles, tied for first in the high jump and took second in the shot put. When I joined the senior team, I was the best in the 120-yard high hurdles, 200-yard low hurdles and second best in the high jump (behind Jack Dalton), shot-put, and discus throw. In one track meet I took first place in five events, a school record.

My basketball skills were not as polished. I lettered but played second string. We were the only high school in St. Louis that did not have a football team. I think I would have been a good football player. My track coach was Mr. Ward Parker, and I hoped that he would help me obtain an athletic scholarship to Missouri State University. I had started smoking when I was 15 and I knew that that would be an impediment to the scholarship.

I never made a grade below "C," had very few "A's" and only a few more "B's." My best grades were in math, physics and chemistry because I liked the subjects and could easily understand the work. All of my teachers would counsel me and tell me that they knew I could do better, but just wouldn't work hard enough. They all knew I worked after school and all through the summer and they would let me take my exams a few days early to leave on my sealing expeditions. They also knew that I should do better even with my outside work.

When I was a senior, I began to run around with a group of young men from our neighborhood who had been drafted during World War II and discharged when the war ended after serving one or two years. All of them enrolled in college using the GI Bill of Rights. Jack Dalton had graduated and was attending Washington University as a scholarship student and studied hard to keep his scholarship, so I started hanging around with the ex-GIs. None of us were old enough to drink, but I do not remember any of the bartenders in our neighborhood checking our age. I looked older and also had worked for some time already and appeared more mature than I was.

On the nights I was not working, we hung around the neighborhood bars on Delmar, drinking beer, swapping stories and enjoying big band and Dixieland music. My new friends were Eldon Brewer, just out of the Navy; Tex Laurence, a Battle of the Bulge Veteran; Bob Walters, Army; and Ed Krummenacher, Navy. Tex was the only one who had been in combat. I listened to all of their stories about boot camp, barracks and sergeants and decided that I would like the military. None of them liked the military and they were glad to be out, but every one of them agreed that it was the best thing that had ever happened to them. They were better disciplined, better focused and had the GI bill for college, which they probably would never been able to attend without the war and the draft

Late in my senior year I still had not received an athletic scholarship. Then I heard about a competitive exam to become a Naval Aviator. The Holloway Plan started out with two years of college as an Apprentice Seaman with all expenses paid and a salary of $50/month. Then you would go into Navy Pilot Training as a Midshipman, US Navy. If you won your wings, you would be awarded a regular commission, serve four years as a pilot in the fleet and then go back to complete your last two years of college as an active duty officer with full pay and allowances. I immediately applied and took a battery of very difficult tests. In about a month I received a telegram from my congressman stating that I was one of two selected from the state of Missouri and the Navy scheduled me for a flight physical. I was on top of the world—selected for pilot training and being awarded a full scholarship in spite of my mediocre grades!

My hopes were soon shattered. I failed my flight physical because the breathing channels in my nose were partially obstructed by cartilage that had been shifted by a broken nose that had not been straightened out.

I was down but not out. I went to my family doctor and he referred me to another doctor who said he could reset my nose and remove sufficient cartilage to open up the breathing channels. I paid for it, despite the fact that there was no assurance that the Navy would restore the scholarship even if the operation worked. The doctor only charged me $300. He did the operation in a dentist chair using a local anesthetic, and I went home. I remember the breaking of my nose and the chiseling of the cartilage. I was in terrible pain as soon as the anesthetic wore off, and I took pain pills for several days. My whole face was swollen and my eyes looked out through slits. My doctor took the packing out and the bandages off and told me the passages were now open. I called the Navy and they reluctantly agreed to look at my nose again. After a lot of consideration by the Navy, they accepted me again. I believe it was because I had a unique job record that revealed a love of the sea and adventure and had demonstrated my determination by fixing my nose myself. It sure was not for my good looks. My face was still swollen and my nose was big enough to suck in all of the air in the room. I never felt at ease in a dental chair again and it took all of the courage I could muster even to get my teeth cleaned. After my nose completely healed I could push my nose all over my face without much interference from cartilage or bone. The Navy also gave me permission to go on my last sealing expedition and report for duty at my college in early September. I missed my high school graduation because I had already left for Alaska, but it was worth it—I was going to be a pilot! Little did I know that the path would be longer than I had planned.

I was 18 years old on my way into the USN as a pilot and going to college with a full scholarship. I figured I had grown up and was really on my own. I knew I would always have the love and support of my extended family, but from now on I would be the master of my destiny.

My mother went to work full time, free for the first time from the worry of raising her children.

~ CHAPTER 3 ~

College Life

My US Navy scholarship covered all books and tuition for any college where I could gain admittance and $50.00 per month salary as an apprentice seaman USN. I knew my grades were not spectacular, and all of my ex-GI friends and several school chums had chosen James Milliken University in Decatur, Illinois. It was a small Presbyterian college with about 1200 students in a small town. I liked small. I was admitted along with Tex Laurence, Ed Krummenacher, Bob Hesse, Bob Tully and Roy Sanderson.

The Navy required me to successfully complete sixty semester hours, to include: two years of English, two years of math including calculus, one year of chemistry, one year of physics and one year of a language. I submitted my plan to the Navy including these courses and a variety of liberal arts courses; e.g. sociology, psychology, anthropology, world history and world religions. Right after I returned from Alaska in early September 1947, I enlisted in the USN as an Apprentice Seaman and shortly thereafter went off to James Milliken University in Decatur, Illinois, with five of my friends.

Bob Hesse, Tex Laurence, Ed Krummenacher and I rented two upstairs bedrooms in a home owned by a very nice elderly couple. We ate most of our meals in the corner restaurants (The Pub and the Blue Mill) near our school. I adapted easily to college life and became more interested in learning.

My very rudimentary learning process at the time was to take good notes and study for exams. Much later I learned that you have to focus on the learning objectives for each course, develop a regular study habit and exploit all of your sensory channels to retain and master the subject matter in your brain. I will explain my current learning process as it evolved during my lifetime.

The professors usually clearly state the learning objectives for the course and each portion of the course. This is what they want you to learn from the course. They also will focus their lectures on these objectives. Finally, they will determine how well you mastered these objectives by subjectively evaluating your written assignments, your classroom responses and your periodic tests. (This is also an excellent recipe for you to follow in living your life or running your business: Establish your own goals with supporting objectives, make a plan to achieve them and execute your plan.)

The Learning Process

My methodology to satisfy learning objectives evolved over my seven years of college and four years of professional and military schooling. First and foremost I focused on the objectives and would always take good notes and listen carefully when subjects related to the objectives were discussed or assigned.

Second, I believe that your brain will learn and retain subject matter that is frequently repeated, comes from several different sources, enters your brain via multiple paths and is validated by testing. The more difficult the subject matter, the greater the need for this multiple-path, repetitive process:

You see it when you read (study) the assigned lesson from your textbook. This is one author's interpretation. You hear your professor's lecture, which is his interpretation. This provides you with his "foot stompers" and should prioritize your focus and allocation of study time. When you take notes during the classroom lecture, you translate your interpretations by writing your summary classroom notes. Then you see it again in your notes and textbook when you study for the exam and see it and write it again when you take the test. Finally, you see and correct your mistakes and reinforce your correct approaches when the test is scored and debriefed. That makes six repetitions through three different sensory paths to your brain of three interpretations of each subject with a verification and correction by test. Even the slowest brains will learn and retain the subject matter. I recommend this process to all readers. (Footstomper).

When I returned to college eight years later, I had learned this process and earned mostly A's, with a few B's, to obtain a Master of Science degree in

Electrical Engineering (compared to the O's and B's, with a few A's, in Liberal Arts at Milliken University). The real problem, which you will discover as you read this chapter, was that while I was at Millikin I was still lazy and more interested in friends and social activities—as long as I could maintain my scholarship status, which was necessary to become a fighter pilot. My real goal was to become a fighter pilot, and the scholarship was the enabler.

After a few months I had gained enough confidence in my scholastic performance that I knew I would not have any problems handling the required workload of 15 credit hours per semester. I also knew that I could handle college-level math, English and science. After all of the stories I had heard about college courses being so much tougher, I had worried that I might have difficulty making the required grades to retain my scholarship. Actually, I was making better grades than I had in high school and was enjoying the classes a lot more. I had always hated English, but my freshman English professor was a real professional. I wrote my term paper on my sealing experiences in the Pribilof Islands, and he awarded me an A+ on the paper and an A for the course. I enjoyed the faster pace and greater depth of coverage in all of my subjects and began, for the first time, to learn more efficiently.

Guys, Athletics, Gals and Beer

Now I felt I could handle an expanded social life, so I made the Millikin track team, pledged the social fraternity with most of the athletes, began drinking beer with my St. Louis buddies and dating college girls. I was enjoying life to the fullest.

I won my letter in Track easily. I was the team's best high-jumper, and second best in the 200-yard low hurdles and 120-yard high hurdles. Occasionally, I took a second or third place in the shot put.

After I had successfully completed my freshman pledge commitment, I had to drop out of the fraternity. I could not afford the initiation fee and dues, and Bob Hesse (my roommate) was the only one of our St. Louis group that belonged to the fraternity. Bob met Jean, the love of his life, in his first year at Millikin and only had time for study and Jean. He married her after they both graduated, had several children and lived happily ever after in St. Louis, owning and operating his own retail window-dressing business. It was a truly wonderful love story, but it ended too soon—Bob died of a heart attack in his early sixties.

After our first semester, we figured out that we could not afford to live in our boarding house, so Tex Laurence and I shared a room in the old GI barracks on campus. Ed Krumenacher and Bob Tully shared a nearby room in the same barrack. We made friends with several of the local boys—Jim Lukens, Walter Hatfield, Paul Kuhl, Cecil Coleman—and an international student from Iraq, Ed Gnome. All of us were fun-loving beer drinkers and fairly ordinary, except for Jim and Ed. We called Jim "Tiny," because he was well over 6-feet tall and must have weighed close to 300 pounds. He was very "country" and loved his bib overalls, which he would wear to accentuate his country ways. He actually was from a medium-sized city and had worked for his father selling General Tires, but he pulled off the country act so well that he adopted that identity. We had a lot of blind dates, and if Tiny found out he had a loser beforehand, he would wear his bib overalls and turn her off proper in the first hour! He had a terrific sense of humor and everyone liked him.

Cecil Coleman had a unique talent in that after he had several beers he could stand on the sidewalk and urinate as high as the windows on the second floor. Thus he earned our immediate respect. We pondered how best to exploit this rare gift, but the only application we ever found for this unique skill was to win bar bets with any (male) strangers that frequented our college pub. I never again have found or heard of anyone else equally skilled. However, I have had little opportunity or inclination to seek others out.

Ed Gnome, from Iraq, probably was from a wealthy family but was a lost soul in Decatur, Illinois. He was very serious and intense but seldom understood our American humor. We found that humor is culturally unique and seldom translates. He listened a lot and seemed to enjoy our company. We enjoyed Ed and made him into a passable American-Mesopotamian. I hope that Ed Gnome faired well in life. We never heard from him again after college.

We had very little spending money, but Paul Kuhl and Walt Hatfield had cars. We would all obtain dates and buy a case of beer. One of the town boys would bring one hot dog, one bun and some mustard. We would then journey out to the local lake in a wooded area for a picnic. Sometimes we would have to make two trips to carry all of the people. We would then build a small bonfire, cook our sacrificial hot dog and drink our beer. I never knew who ate the hot dog—probably Tiny Lukens. We would enjoy our dates, sing a little, neck a lot and eventually go back to the barracks—alone. Most of the time we would stay after dinner at the local pub, drink beer, tell Army (Tex), Navy (Ed), Marine

(Bob Tully) war stories and discuss politics, girls and future plans.

I dated a lot of really neat girls and enjoyed all of them. I remember Barbara Hall and Carolyn Duncan, but the rest of them are just smiling faces. None of us became serious. All of us, except Bob Hesse, married girls that we met later after college. I had no intention of looking for my future bride until I was established as a pilot and could support a family.

Bob: 20 years old

The US Navy allowed me to go off on my third and last Sealing expedition during the summer vacation in 1948. On my way home from Alaska, I stopped off in Medford, Oregon, to meet my new brother-in-law's family. His dad was involved in a wildcat oil well and I invested $400 of my summer earnings in the oil venture. I had visions of wealth; but, alas, I lost all of it! This was the first (but not the last) of several get-rich-quick investments during my lifetime. As a result, I could no longer afford to finish out my two years at Millikin. Fortunately, the US Navy allowed me to transfer to Washington University back home in St. Louis for my last semester. I lived at home, dated some of my high-school girl friends when they came home for weekends, and reported into Pensacola Naval Air Station on 15 June 1949 for US Navy Pre-Flight. It was the beginning of my 30-year military career and 27-year civilian career. (A college classmate from Milliken, Walter Hatfield, had enlisted as a Naval Aviation Cadet and reported into Pre-Flight six months earlier.)

As a child, I had built model airplanes and watched the planes at Lambert field. I loved military war movies, longed to fight for my country and always wanted to fly. Thus, I was very pleased to be in Pre-Flight and felt that flying would play a significant role in my life. Flying was also much more attractive than being a seal blubberer or a gandy dancer on the railroad.

When I was very young, I had pursued adventure and travel and needed an income. I satisfied these needs and desires by using the only skills that God had given me. I was always tall and strong and could qualify for jobs in exotic places, particularly during WWII, when the normal labor pool was engaged in fighting a war. Little did I know that my days of hard labor would also motivate me to become an executive who works with his brain instead of his body. I recommend hard labor for unfocused young people. It is a powerful motivator to find a more challenging lifework!

~ CHAPTER 4 ~

Transition to the US Navy

There were about 50 of us who reported to duty as Naval Aviation Class 15-49 at Pensacola Naval Air Station Florida. The US Navy Holloway Plan had expired and most of our class were the last group of Midshipmen on the tail end of this WWII program. Each of us would be sworn in as Midshipman USN, the same rank as the US Naval Academy students. The only trouble was that public law read that duty as a Midshipman USN would not count for time in service or toward retirement—appropriate for Naval Academy students attending Annapolis, but not appropriate for active-duty troops in flight training. Our other classmates held the rank of Naval Aviation Cadet or Ensign, and their time counted for active-duty service and retirement. The Naval Aviation Cadets also were commissioned Ensigns as soon as they won their wings (14 to 19 months). A Midshipman held his rank for two years. Thus, if he was awarded wings in 18 months, he went out to the fleet as a Flying Midshipman until he was commissioned as an Ensign at the two-year point. As a result, when the Korean War began we had Midshipmen flying combat missions, and their time did not count for service or for retirement. Congress remedied the situation about 15 years later.

The "Keystone Kop"

When our class arrived and settled into the barracks, a group of us went to the gym and played basketball. My competitive juices must have been flowing, because someone tried to block my shot from behind on a breakaway and we both went down at a fairly fast speed. My leg was broken. I was taken to the US Navy hospital, where I was placed in the Sick Officers Quarters. The problem was that our class had not been sworn in as Midshipmen yet, so I was still an Apprentice Seaman—really in limbo. The hospital pretended I was a midshipman and kept me in officers quarters. Class 15-49 was sworn in the next day and started Pre-Flight. I had my leg set and a cast installed and spent the next seven weeks playing cards with my salty old roommate, a commander from the submarine navy. He taught me how to lose money gracefully, like an officer and gentleman, and more than I needed to know about subs. I grew very fond of him.

My doctor, Doctor Lt./Commander Walters, was the first female US Navy Flight Surgeon, small in stature, but very attractive and talented. I aroused her wrath a few days after my cast had hardened. I went into our bathroom and used the shower curtain bar around the bathtub to do my pull-up exercises. The whole works collapsed and I broke my cast. She replaced the cast and chewed me out for overestimating the structural integrity of the flimsy shower ring. She had a valid point and made a convincing case. One week later, after my second cast was good and hard, I found a young ensign in a wheelchair and challenged him to a race down the deserted hospital corridor. We started at the end of the corridor and were racing neck-to-neck, when a door suddenly opened into the corridor. It turns out that all doors except the storeroom door open up into the rooms. A Navy corpsman was backing out of the storeroom carrying supplies, when he was flattened by the door, followed by two wheelchairs and their occupants skidding down the corridor with his broken supplies. I had hit the door "cast first" and once again broke it. I was lucky I did not rebreak my leg. They X-rayed my leg and it was fine, but once again I had to face the furious Dr. Walters. After a considerable amount of one-way discussion Dr. Walters again scheduled me to the cast room.

She directed the casting personally. After the old cast was cut away, the corpsmen began applying the wet plaster strips. They went on and on, and when they were finished my cast was at least twice the size of the old cast and heavy as cement. She surveyed the completed job with this evil look, and told the corpsmen that the cast was so large that I would be unable to take my pajama pants off. She smiled and then told them to cut my pajama pants off and issue me a hospital gown, which I wore during the rest of my tour in the hospital. She also told me that she had many other creative cast designs should I happen to break this cast. I could not leave the bed until the cast was dry, and that took forever! The nurses all made comments on my hospital gown and I was the butt of many jokes in the Sick Officers Quarters until my cast was removed. I then had to have physical therapy to rebuild the strength in my gimpy leg so that I could renter Pre-Flight with its very demanding physical fitness requirements.

I had been in the Navy nine weeks when I was discharged from the hospital after passing my flight physical again. They assigned me to the cadet barracks where I worked out daily, walking and running on the sandy beach to build up my stamina while waiting for another class assignment. I was finally sworn in as a Midshipman USN and assigned to class 18-49. I was the only Midshipman in the class and 3 months behind my original class. I also had never been off the base.

A Tough Lesson

My friend from Millikin, Walter Hatfield, had just completed Carrier Qualification and visited me in my barracks. My Class had not yet started, and I decided I would sneak out of the barracks and celebrate Walt's successful completion of primary training in downtown Pensacola. We put on civilian clothes and went into town in his car. After a few beers we started back to the base, and a car pulled alongside us and told us to pull over. It was dark and I did not realize they were police in plain clothes. They seemed to be giving Walt a hard time, so I got out of the car and came around to back up Walt. They became very upset. At that point I realized they were police, but it was too late. They took us down to the police station and turned us over to the shore patrol. I had blown it again. I was absent without leave (AWOL). Walt received a verbal reprimand and went off to Corpus Christie for advanced flight training. I was scheduled to meet an elimination board to consider termination from the USN because of the AWOL charge. It was a very low point in my life, and I had only myself to blame.

The board met and I testified, confessing that I had gone off the base without permission to celebrate with my friend. The Board awarded me the maximum number of demerits allowed in Pre-Flight training (50) and allowed me to stay in Class 18-49, which was due to start the following week. That meant that I had to complete the 16-week Pre-Flight course without receiving a single demerit and walk penalty tours every weekend and holiday. An unshined shoe, a dusty shelf, a missed salute would result in my elimination from Naval Flight Training.

Bob: In Flight Training

Our Pre-Flight Tactical Training Officer was Captain Erskine B. Crew USMC, and he advised me that he would be watching me every minute, looking for an excuse to give me that single demerit that would eliminate me from the Navy. Well I gave it my very best effort for the first time in my lazy life. I excelled in the academics, the military bearing and the athletics. I was the top boxer, wrestler and swimmer in our class and received a perfect score of 4.0 in athletics. I made the USN Pre-Flight Drill team, and we drilled during the halftime at college football games. (On the college road trips I was always confined to quarters.) Captain Crew watched me like a hawk, inspecting my cubicle and myself at every opportunity. I made it through the 16 weeks of Pre-Flight without receiving a single demerit! After our graduation Captain Crew congratulated me, wished me well and told me that he had applied maximum pressure to eliminate me but I had stuck it out. He made me feel very good. I even wanted to become a Marine pilot. Our paths were to cross again a few years later.

At last, I finally was allowed off base (legally) on a weekend pass after 26 weeks in the Navy. I wore my uniform, which was required, and behaved like an officer and gentleman. The gandy dancer and seal blubberer in me were suppressed ... but not quite dead.

The Joy of Flight

All of our Class 18-49 were transferred from the main US Navy base in Pensacola to an outlying auxiliary US Navy base— Whiting Field in Milton, Florida—where we would learn to fly the SNJ single-engine trainer.

At last I began flight training and loved every minute. I had never driven an automobile because we never had one, so this was my first interface with a vehicle larger than a bicycle. My instructor was very good and I rapidly overcame the initial difficulties of first under-controlling this large mass of inertia, then over-controlling the stubborn beast, and finally oscillating about the point of optimum control until I settled into well-trimmed flight. My instructor was almost as pleased as I, when I soloed. It was wonderful to be alone in control of this remarkable flying machine. That night I wrote home and proudly announced the momentous event. My mother, grandmother and doting aunts were all very pleased and proud.

The next phase was acrobatics, with and without an instructor. The instructor would demo some acrobatics; then, when I was reasonably proficient, I would be scheduled for solo practice flights to build up flying time and confidence in acrobatics and landings. I loved acrobatics and did very well. I also felt very confident on my landings.

A Sobering Lesson

The SNJ had only two landing gear and a tail wheel that was either locked (for takeoff and landings) or unlocked for taxiing. We could not see ahead of the aircraft while taxiing, because the cowling blocked our view. So, we made "S" turns whenever we taxied, to avoid running over obstacles or people walking in the taxiways. We also learned never to wear our helmets while walking on the taxiways going to and from our parked aircraft, so that we could always hear an aircraft approaching. I was taxiing to my parking spot after one of my solo acrobatic flights, when the tower operator told all of the aircraft taxiing to stop and hold position. I stopped and scanned the field to see what was wrong. The SNJ just ahead of me had S-turned into an ensign student-pilot who had been walking diagonally across the taxiway carrying his parachute with his helmet on. The propeller had killed him instantly, and there was a gory mess of man and parachute on the taxiway ahead of me. The student pilot who had killed him never saw him because his victim's walking path was blocked out by the blind spot and his S-turn tracked the victim. The victim never heard the SNJ because his helmet was on and never saw the SNJ because it approached from behind his right shoulder. It was my first experience with an aircraft fatality. It was very graphic and illustrated the dangers surrounding aviation and the need for attention to detail in everyday operations, however small it was a very sobering lesson that I never forgot.

We also had to qualify for night flying. We had a lot of classroom instruction on the rod and cone sensors in our eyes. The cones were our high-resolution day sensors with very high resolution in the center of each eye and ever declining resolution as you proceeded out from the center. The rods were our night sensors and also very good motion detectors. When your rods detect motion anywhere in your field of view, they immediately focus on the moving object and identify it, using the very high-resolution cone sensors in the center of each eye. The rods were the lower resolution night sensors and, like the cones, the resolution declined as you proceeded out from the center. The rods were blinded by white light, and you had very little night vision when the eye was exposed to white light. Red light, however, did not degrade night vision. Our classroom instructor demonstrated this by exposing us to white light and then closing us in a very dark room. He placed an eye chart on the wall and asked us to read it as soon as we could see the numbers. At the end of a few minutes most of us could read the big E. As time went on, we could read more lines; and soon we could see the details in the room pretty well. He then illuminated an aircraft instrument panel with very faint red light, and we all could read the instruments and still see the room details. Then he rapidly turned the white lights on and off. We were as blind as bats and had to turn the red lights up until our night vision was restored. It was a very convincing demonstration.

Within the next few nights we had several night flights with our flight instructor. He showed us how to take off with our red lights bright enough to read the instruments and then gradually turn the red lights dimmer and dimmer until our night vision was established and we could see the landscape even on a moonless night. Then we started night practice landings without using the white landing lights installed on our wings. When we'd made several good landings, the instructor climbed out and we then demonstrated three night landings all by ourselves. After a few more solo night flights we planned and flew a round robin navigation flight at night, and our instructor checked our navigation log to see how closely we had come to our time-and-fuel estimates at our three checkpoints. We were now halfway through primary flight school and were transferred to Corry Field in Pensacola for instrument flying.

Thunderstruck

On the first Friday back in Pensacola, my friend Owen Dykema and I went over to the Gadet Recreation Center for the weekly dance. Young ladies from Pensacola would come out to the Main Base and meet and dance with Cadets and Midshipmen. College vacation had just started and we wanted to check out the new crop home from college. I was not inspired and was ready to leave when Owen came up and stated that he had just seen a living doll! He pointed out a group of ladies all gathered in one spot. I told him that I had already looked them over and was ready to leave. He asked me if I had seen the girl sitting on the couch surrounded and masked by the other girls. I went over and had to stand on my toes to even see her. I was THUNDERSTRUCK.

I asked her to come out and dance with me. As soon as she emerged and smiled at me with those big sparkling eyes, I knew I was hooked. Owen was engaged, but he had spotted the girl of my dreams and he earned my undying gratitude for providing me with the lady that made my life very happy. We danced until she had to leave; then I accompanied her and her two best friends out to her father's stretch limousine for the journey home. She explained that she was the second oldest of eight children and it took a stretch limo to take the family anywhere. She was very shy, but I could tell that she liked me immediately. I was very tall and she was tall, and that was good. She had attended Catholic schools and was a sophomore at St. Mary's Dominican College in New Orleans (an all-girls school). She also had to be home before midnight, and that was good. She dropped off her two friends and then drove to her very large home. I took her to her door and asked to see her again. She invited me to come over the next day and meet her family. I felt that I had found the love of my life, and I walked the four miles back to the base barely touching the ground. This was a very lovely unspoiled lady with a beautiful soul, tailor-made just for me.

Calamity Strikes

I started instrument ground school at Corry Field the next week and had a very hard time concentrating on anything other than Mary Sabra Sanchez. I was seriously sidetracked and very happy. I was learning to fly and had just met the lady I hoped to marry.

The Navy was also THUNDERSTRUCK. It was June 1950 and the US Navy was in a major crisis. President Harry Truman's Secretary of Defense was Louis Johnson. For years there was an ongoing debate on the need for building B-36s for the USAF or more Aircraft Carriers for the US Navy. The decision was made to buy the B-36s and to drastically reduce the military budget and the number of military on active duty. The US Navy's budget was cut drastically, and a large reduction in force size was being executed during the last few months in the current fiscal year (April-June 1950). Our reserve flight instructors were being released to civilian life, the USMC instructors were reverting to their permanent rank, and several of our instructors were involuntarily reduced to their enlisted grade.

Morale was very low among our instructors. Their workload was increasing as more of them were released from active duty. And for those reduced in rank—pay was being reduced.

More work for less pay! The quality of instruction and quantity of flight instructors were decreasing and the student/instructor relationships were becoming strained. I had not noticed these stresses on the flight instructors at Whiting Field, but it soon became obvious at Cony Field.

My flight instructor at Corry was a real grump, but I was in love and nothing could bother me. He was an enlisted pilot, and I later found out he had been a full Lieutenant USNR who had just recently reverted to his enlisted rank. He also had five students and most of the flight instructors had four. The ground school was very good and we had a lot of time in the Link Trainers (flight simulators). Thus we satisfied the instrument flying requirements in the flight simulators before we had to demonstrate them to our flight instructors in the aircraft. I felt very comfortable flying blind in the simulators and could execute all of the required maneuvers without difficulty. In the aircraft we flew under a hood in the rear seat of the SNJ so that we could not see outside of the aircraft to demonstrate our ability to takeoff, land, fly and navigate safely in the clouds and fog. The instructor flew in the front cockpit and visually kept us clear from other aircraft. He also corrected and scored our performance.

I had been flying most of my instrument practice flights in mornings, without difficulty. My instructor often waited until I was in the more difficult maneuvers to ask me a question. I would have to pick up my hand microphone and answer the question. He explained that this was his way of adding additional workload to see if my performance would degrade while using one hand to hold the mike and talk. He did it a lot, and it irritated me because it delayed my maneuver and disrupted my scan pattern. My final check ride was scheduled in the afternoon, which made it a more difficult experience. We flew below the fair weather cumulus clouds, and the afternoon turbulence was significantly greater. I really had to work hard to execute the required maneuvers in the turbulence. When I started the partial panel maneuvers where you only could use the altimeter, rate of climb, needle and ball and magnetic compass, it became very difficult; but, I was performing within the allowable envelope. Then he started asking me questions and I got behind the time envelope for the maneuvers. He failed me on the check-ride—my first flying failure. We rescheduled the ride for the next afternoon. It was very turbulent again and we stayed below the cumulus cloud level, where the turbulence is maximum. Once again he asked questions during the partial panel maneuvers, and once again I took too long to complete the maneuver. He failed me again and scheduled me to fly with the flight leader for a final check. I had to pass this one or be eliminated.

A few days later I was scheduled to fly my check ride. I was greatly relieved that I was not flying with my instructor, and felt I could easily pass with any other instructor. My morning check ride was then changed to mid afternoon ... and I figured then that I was about to repeat the previous experience but with a different instructor, who was about to repeat my instructor's training techniques. Sure enough, we stayed below the cloud level in maximum turbulence and the check pilot asked me questions during the partial panel work. I completed the first partial panel series within the envelope and thought I had it made. Then he had me perform a second set and asked more questions, and this time I missed the time envelope. I was very upset and felt certain that I was selected for elimination. We returned to base and landed ... and I was subsequently eliminated from the program. It was the greatest disappointment of my life and the end of my naval career.

Within the week, North Korea invaded South Korea. President Truman's Secretary of State Acheson had given his famous perimeter speech, where he outlined the US "line in the sand" where we would not allow communism to expand. Unfortunately, he had omitted the Korean Peninsula and shortly after his speech North Korea invaded South Korea. Equally unfortunate, our Secretary of Defense had already drastically reduced the military budget and manpower, and the US was now inadequately manned to deal with this diplomatic blunder. It was the first of many blunders of Democratic administrations that resulted in an undeclared war. The reserves, many of them just involuntarily released from the service had to be recalled and the military was scrambling hard to respond to the crisis.

The US Navy was in particularly poor shape. I was still in the barracks awaiting discharge when the war started and tried to get back into the program but received no consideration. I vowed that I would fight in that war if I had to go to Canada to do it. I knew that I could fly with the best of them and was determined to prove it.

Looking back on that sad event some 52 years ago, it was the best thing that could have happened to me. I would never have completed a career in Naval Aviation. Like many of my US Navy classmates, I would eventually have had to choose between my family and the US Navy. The excessive amount of sea duty and the number of six-month cruises that turned into nine-month cruises were very stressful on family life. Many of my Navy classmates left the Navy at the 8-to-12-year point, mainly because of family stress and sea duty. I probably would have separated from the Navy about the same time and gone for a civilian career or the airlines. In retrospect, my very short Navy career was a very lucky calamity; and, best of all, I had met the Lady that I wanted to marry.

Now I will tell you about Sabra's family and her life before she met me.

CHAPTER 5

Mary Sabra Sanchez
Her Family

Mary Sabra Sanchez was born on May 25, 1930, in Pensacola Florida. She was the second of eight children born to Phillip Joseph Sanchez II and Mary Sabra Quina Sanchez. The third child was a boy, Phillip Joseph Sanchez III and all of the other children were girls.

Both of her parents came from old Pensacola families. Her mother was a descendant of Desiderio Quina (Italian spelling— Chini), who was born in Bologna Italy in 1777, immigrated to America about 1800, married Margerita Bobe of Pensacola in 1813 and died in 1830. He, his wife and many of their descendants are buried in St. Michael's Cemetery. My wife was the fifth generation of Quinas in America. Other ancestors included the Bonifays (Martinique), Caros (Spain), Golays (Switzerland), Bobes (France) and Laniers (England). Her mother was one of seven children.

Her father, Phillip Joseph Sanchez II, was the grandson of Francisco (Frank) Peter Sanchez born in San Antonio, Texas, in 1836. He married Margaret Mary DeVoy (from Cork Ireland) in Mobile Alabama in 1859. They had five children, and he fought in the Civil War as a Confederate artillery officer in the 1st Alabama Company C. After the war, they moved from Mobile to Old Warrington, Florida, where they opened a Ship's Store on what is now the

Naval Air Station. They owned and operated the store for 24 years, until Frank died in 1877. Their youngest son, Phillip Peter Sanchez, married Mary Margaret O'Donovan, a Protestant, in 1896 in the parlor of St Joseph's Catholic Church in Mobile Alabama. They were my wife's paternal grandfather and grandmother.

Mary Margaret O'Donovan was the first of eight children born to John O'Donovan of Limerick, Ireland, and Moline Whitehead of Newport, Florida. They were married in Christ Church at Pensacola, Florida, in 1876, and John O'Donovan became a US citizen in 1880.

Mary Margaret and Phillip Peter Sanchez had 13 children. Phillip Joseph Sanchez II was their sixth child. He married Mary Sabra Quina on May 7, 1927, in Pensacola, Florida; and I married his second child, Mary Sabra Sanchez. As you can now understand, she was related to hundreds of people in Pensacola. Rueben Askew, a two-term Governor of Florida and Democratic contender for President of the US, is one of her many cousins. More than 300 people attended a recent family reunion on the Sanchez-O'Donovan (her father's parents) side of her family. It is reasonable to believe that an additional 300 could have attended, if the Quina family had been included.

My wife's family, to me, was like an American fairy tale. They were very close to one another, were educated and raised in the Catholic church/school community, and love and faith abounded. Sabra's mother was a devout Catholic and devoted mother who loved her husband, her sisters and her children, and seemed to like me well enough to allow me to date her daughter. Sabra's father was clearly the head of the family, very protective, and from the beginning was wary of me as a suitable boyfriend for his favorite daughter. (All of his daughters thought they were his favorite daughter, which is commendable). It was very clear that Phillip Joseph Sanchez III, Sabra's younger brother, was the heir apparent and was expected to follow in the footsteps of his father. (His father seemed to be critical of his progress.) It was also apparent that Sabra's father viewed me as a high-risk for marriage.

I had a lot in common with Sabra's father. He had started working in the San Carlos Hotel as a shoeshine boy when he was very young. He stayed with the hotel and eventually became a bellboy, then a hotel clerk and finally the manager. He dropped out of school early and helped pay for his older brother's (Lars) education and later went into a partnership with him and another friend. His family was poor and all 12 children had to work and help one another. Phil continued to help his brothers and sisters, even after he was married and had his own large family to support.

When WWII started, Phil was the assistant manager of the San Carlos Hotel. The family who owned it placed their son in his position to obtain a draft deferral; so Phil went into the insurance business, hiring on with the Equitable Life Assurance Company. By this time he had acquired several apartment buildings in the Pensacola area and was renting them to the many US Navy officers in flight training. The war and mobilization were underway by now, and the terrible depression was finally ended. Apartments were hard to find, and the US Navy student pilots coming to Pensacola to learn to fly were greeted by "no vacancy" and "no children allowed" signs. Phil Sanchez was featured in the New York Times as a rare landlord who did not raise his rents and refused to rent to young couples unless they had children. He also held barbecue picnics for his tenants on Sundays. Nearly all of his tenants were training to become US Navy pilots. Gene Tierney's brother and Van Johnson were among his many tenants. This selfless act was typical of his business ethics and his strong family values. Patriotism also played a large role.

After WWII ended, Phil converted five of his apartments on Sanders Beach into the Bayou Vista Hotel. He had always loved the hotel business, and this was his way to get back into his desired profession. The hotel was a block away from his home on 1421 Cypress Street, and he used his children to run the switchboard and manage the desk. He continued selling insurance full time with Equitable and was always a member of the million-dollar club. Throughout his life he continued his self-education— studying law and obtaining his Certified Life Underwriters rating. He served as a member of the Santa Rosa Island Authority and as Commodore of the Pensacola Yacht Club. He was a self-made and successful businessman, well respected in his community and well loved by his family.

Sabra's mother was the queen of the house. She had a nanny to assist her with the family and an elderly black handyman. George had been with the family forever and was kind of a retired yardman. His chores were few and far between, but he was very good company and enjoyed all of the children. George was a very good friend to Sabra's father, and they often talked. George was also a retired preacher. He stayed working with the family until he died. The entire family sorely missed him.

Sabra's mother was an excellent cook but always said that she hated cooking. She taught all of her children to cook and evaded the kitchen whenever she could. She was very close to her younger sister, Rachael, who lived one block away with her husband and six children. The two families with their 14 children were inseparable and they all grew up playing and attending the Catholic schools together.

Sabra's mother and two of her mother's sisters had gone into the convent at an early age. Sabra's mother dropped out, and the other two sisters became nuns and lived in Mobile. One of her sisters died of an overdose of radiation from a routine lung X-ray before I came into the family. Her other sister, called Sister Terisita, taught school in Mobile for her entire life. We called her Aunt Wissie. All of them and their children worshipped their mother, Grandma Quina, who also lived nearby. All of the Quinas were devout Catholics with a very deep faith. My wife Sabra inherited this deep faith, and it provided the anchor for our wonderful marriage. Her faith eventually offset my prejudice toward the Catholic religion. When you marry, you marry the whole family; and I was truly blessed by entering this family.

Mary Sabra's Childhood

Mary Sabra was a bashful and very affectionate child, who loved to sit in laps and be cuddled. She spoke very softly during her entire life, probably because she was such a bashful child.

She also was near-sighted and would study your face carefully with her sparkling big brown eyes, which devastated me completely.

Others perceived her to be "stuck up" because she would not acknowledge those who waved or spoke to her from across the street. Actually, she could not recognize people from a distance, and that also worked in my favor. I stayed very close to her, where she could see me very clearly, and where I could treasure her every word. I held her hand when we walked, and I put my arm around her when we were seated. I sure did not want this Lady to get away.

Dolores, Phillip Joe, Sabra and Annie

When she was in St. Stephen's grade school, she had a wonderful young nun as a teacher, Sister Mary Wilfrid. The classes were small—

perhaps 12 students—and Sister Mary Wilfrid recognized that Sabra's bashfulness was impeding her scholastic and social progress. She commenced a very wise long-term program to overcome her bashfulness. She placed Sabra next to her at her desk and would occasionally ask her to answer questions or recite, encouraging her to speak loudly enough so the rest of the class could hear her. She would also occasionally place her arm around Sabra's shoulder. When Sabra became less bashful, Sister Mary Wilfrid placed her back at a desk with the class and gave her bit parts in school plays or required her to read or recite poetry to the class. Sabra gradually became more social and lost much of her bashfulness, largely because of the efforts of Sister Mary Wilfrid. Sabra remembered her and loved her for her entire life. When Sabra died, Sister Mary Wilfrid was in a retirement facility for nuns in Cullman, Alabama. She was 93, sharp as a tack and remembered Sabra very well as one of her dearest students. The nuns were the heart and soul of the excellent Catholic school system, and they are now few and far between. They, too, are sorely missed.

Sabra played a lot with her little brother and his friends and was a bit of a tomboy. Phillip Joe was about 18 months younger and always looked up to and worshipped his two big sisters. Sabra became the informal leader of her brother's boy gang and shared a lot of good times with them until the boys' voices changed. Buster Zanger and Charlie Van were two of the culprits.

My favorite story about young Phillip Joe happened when he was about 16. He and Charlie Van had acquired an old "Pirates Treasure Map." After diligent study, they figured that the most likely location of the buried treasure was under south Palofax Street just south of Trader John's Bar. They brought a pickax and two shovels, borrowed a yellow "Men at Work" sign from the city and commenced work. They had penetrated the street surface and had dug several feet, when they figured that they were attracting too much attention. So they gathered up their tools and took a very extended coffee break. They were not caught, but it was years before Phillip Joe told us the story. It was a good story, and I still enjoy it. I never learned what happened to the treasure map. I would like a try at the treasure. Perhaps the treasure is the memory.

The High School Years

Sabra then attended the new Catholic high school, which had opened its doors three years earlier. The dominant nun was Sister Consuella, and she molded her students into ladies, gentlemen and scholars ... or else!

Sabra made many good friends outside of her own family—Ann Wilson Smith, Catherine Leary Humphreys and Geraldine Foss Barker—all of whom she stayed in contact with until she died. Of course she also had her sisters, brother, and literally dozens of her cousins attending high school with her. She remembered a brand new young priest, Father Cunningham, who would play baseball with the students during recess. Later, we had reason to remember him forever, since he provided us with our first adopted child, Mary Kellie Kennedy.

Sabra's father owned two waterfront blocks on Pensacola Bay just east of Sanders Beach, which extended out into the Bay. When Sabra was seven, the family moved from Garden and E Street (right next to the old hospital and very close to Grandma Quina) into a very large home built on the waterfront about 1908 by the Duval lumber family. It was a large two-story home with an attic and with a large mahogany staircase and trimmings. Aunt Rachael and her family lived two blocks away, and the new house was large enough to hold all of the children. They could still walk to St. Stephen's Catholic grade school and both Grandma Sanchez and Grandma Quina lived along the way. It was an idyllic life in a paradise setting, and family problems were few and far between. Sabra's father handled the few disciplinary chores and made all of the decisions. The girls were expected to go to college, marry well and have lots of children. Phillip Joe was expected to follow in his father's footsteps.

Weather forecasting was very poor during the 1930s, and Sabra's parents took a brief out-of-town trip, leaving all of the children with "Aunt" Dorma, their nanny, in the big home on the bay. A storm came across the Bay while they were gone, and there was no power or telephones for days. Sabra's parents were frantic. Aunt Dorma was understandably scared; the children were in awe but unharmed. Several years after the storm, Sabra's father had the entire house jacked up and moved several hundred feet from the bay to the corner of E and Cypress. They still had a clear path to the Bay, but there was an intervening street. Sabra's mother had always worried about the children being so close to the water and now she was more at ease—and a block closer to her sister. Phil had a greenhouse built next to their home, where she could raise her beloved African Violets and they all settled down in their wonderful home in the new location.

Phil also bought a stretch limousine, in those days built for funeral homes. He was very proud of his family and enjoyed taking his entire family and clients or daughter's suitors to church or dinner in his large limousine, thoroughly enjoying the inevitable stares of onlookers.

He had owned a large Packard convertible before the family outgrew the vehicle. His wild Irish mother, Grandma Sanchez, loved to take rides in the old Packard. As soon as Sabra and her sister, Annie, obtained drivers licenses, they would be assigned that chore. Grandma sat in the rear seat and always wanted to ride up and down Palofox and Garden streets, which were the main part of Downtown Pensacola. When Grandma Sanchez' son had managed the San Carlos Hotel, he would put her in a corner room on the first floor at the Palofox and Garden Intersection, and she would sit out on the small balcony and watch the people. By this time Annie was about 18 and Sabra 16, and the downtown streets always contained sailors in uniforms. Annie and Sabra were both very beautiful and Grandma Sanchez was very proud of them (and all of her Grandchildren). She would scoot down in the back seat of the convertible as they approached a pair of sailors, place her two fingers in her mouth and loudly blow the mating whistle, pfffrt phhhew. Annie and Sabra would blush beet red, Grandma Sanchez could be heard chuckling, hidden in the back seat and the sailors probably had a very brief thrill.

Annie and Tex

Annie fell in love with Kenneth Wayne Atkinson, "Tex," when he was sent to Pensacola as a Midshipman USN to become a Naval Aviator. Tex was almost two years ahead of me in flight training and was out in the fleet as a flying midshipman when I met Sabra. Annie had traveled to California to see Tex before he went to sea and they were married in California in August 1950, six weeks after I had met Sabra. The Korean War had started and Tex was soon flying combat missions in Korea. I had never met Tex and did not meet him until one week before Sabra and I were married (more than two years later). We were to become very good friends for the next 50+ years. Annie and Tex were blessed with four boys and two girls while Tex was in the Navy. Tex left the Navy after 12 years of active duty, much of it at sea, and joined the naval reserve, flying out of New Orleans until he qualified for a 20-year retirement. He went into the insurance business with Sabra's father and brother until he formed his own General Agency. They lived and worked in Pensacola and Orlando and finally settled down in their home in Gulf Breeze, Florida, in 2001.

This was Sabra's life and family when I met her on June 15, 1950. All of the family was still living in the big home on Cypress Street. Annie was pining for Tex and soon to leave to visit him in California. Sabra was home on summer vacation from Saint Mary's Dominican College in New Orleans after completing two years of college. I was a lovesick Midshipman soon to become a civilian with an unknown future. Sabra's mother and father had just lost their oldest daughter to a Navy pilot and were concerned over Sabra falling in love with a non-Catholic, seal blubberer. Our wondrous courtship had begun!

CHAPTER 6

Our Courtship

Sabra's Father Phillip Sanchez

Sabra and I fell in love with each other on the first night that we met. We talked and laughed and found that we had more in common every day. Our past lives had been drastically different—her values and faith were much stronger than mine, but they inspired me to improve my own. I admired her strong moral character, and I could see the same kind of strength in this still-shy lady that I had seen in my grandmother and mother. I knew that she would eventually make me a better and stronger person, and I wanted her to be the mother of our many children. We had a kind of unconditional love for each other, similar to the unconditional love that I had received from my mother and grandparents. I realized that I had been lonely for most of my life, and with her I would never have to be lonely again. We also were both totally committed to marriage, and I knew that we would never ever break up like all of the marriages in my family. We were clearly made for each other ... forever.

I had to work very hard to prove myself to her father. He was determined to see that Sabra and I had very little time alone together. When I would come to pick Sabra up for a date, he would manage to arrange for us to take care of Sabra's five little sisters. For evening dates we would end up staying home, watching our little ladies; and for beach dates they would always end up coming with us and bringing some of their friends. After a while I invited them to come along because we both enjoyed their company. They had become my little brother and little sisters. Her father would let me take the hotel van when we went to the beach, but not if we went out together at night. Curfews were very tight—10 p.m. on weeknights and midnight on weekends. I bought my first car, a 1940 Ford, while I was still in the Navy just to take Sabra out at night. You guessed it; we still spent most of our evening dates watching the children.

Sabra's Family Christmas
Front Row: Phillip Joe, Terry, Pinky, Cathy, Girlie (Collie Dog) Back Row: Sabra (Mom), Sabra (mine), Delores, Annie, Molly and Phillip Sanchez (Father)

Sabra's father enforced his curfews by locking all of the downstairs doors at the appointed hour so that violators would have to ring the doorbell. We only missed curfew once, and he was waiting for us when we showed up 15 minutes late. I had been having car troubles. He asked me if I ran out of gas (gas gage broken) or had a flat tire, and I answered "No, sir" to both questions. He then stated he wanted to talk to me at 8:00 the following morning. I showed up, but he never said a word. It set our troubled relationship backwards, but we never missed curfew again. Many years later, when we related our story to Annie, she told us how she had dealt with the curfew. One night she was late for curfew and found Phillip Joe outside the locked door—he also was late. Phil told her how he had been beating the system. He climbed up the lattice to the balcony on the second floor outside their parent's bedroom, quietly opened the unlocked door, crawled through their parent's bedroom and went out the bedroom door into the hall without awakening either parent. He then went downstairs and let Annie in. I often wondered if their mother and maybe even their father were quietly chuckling under the covers.

My family had never had a car, so I had never really learned to drive.

Aunt Wissie (the Nun)

I could fly an airplane before I could drive a car, but I had obtained a driver's license using a friend's car before I reported for pilot training. I really learned how to drive using Sabra's father's limousine and the hotel van, hauling the kids around. Sabra's mother and her sister, Aunt Rachael, always liked to be driven in the limousine to Mobile to visit their sister, Sister Terisita. The family all called her Aunt Wissie. Sabra and I would drive them to the nunnery and pick up Aunt Wissie and two or three other nuns, and then chauffeur them around Mobile while they shopped. Sabra and I would sing while we drove and the ladies in the back seat would be very pleased. So I was really courting Sabra, her five little sisters, her little brother, her mother, her two aunts and two or three nuns. I was making progress with the brother, sisters, mother, aunts and nuns but her father was still wary.

Sabra's mother invited me to eat dinner at their house frequently, and I believe it was to increase my exposure to Sabra's father. She was on my side even if I wasn't Catholic. Sabra's father began to talk more to me at the family dinner table and to invite me to eat with the family on Saturday nights at Martines Restaurant and Sunday breakfast in Warrington. He was a very handsome man and a wonderful host. He looked like Governor Dewey of New York and would occasionally be mistaken for him on his trips to New York. He was very proud of his large family and would preside at the head of the table from appetizer to dessert. He was a wonderful role model and I loved the entire family.

When Sabra and I were babysitting for her little sisters, I would always try to work out a scheme so that we could steal a few moments alone to smooch. One time I tied Cathy and Terry together back to back with their pigtails. I gave Pinky an egg through the crack in an open door and gave Molly and Dolores a broom to hold a dinner plate flat against the ceiling. Then, while they were giggling trying to resolve their situations, Sabra and I went into the living room to steal a few kisses. The best I could do with that diversion was two kisses. For another diversion, we would take all of them into Phillip Joe's room, turn out all of the lights and pull the window shades. We would all sit together on the floor in a circle and I would narrate the scariest ghost stories that I could conjure up. I would build a story up to a sudden crisis and then spring a scary surprise. They would all scream and shriek for a minute or two while Sabra and I tried to kiss, but we would both be laughing and would end up kissing teeth!

When we went to the beach or the bluffs on the scenic highway in my car, we would drive with the windows open. I had no horn, and I trained the girls to shout "Beep-Beep" whenever I pressed on my horn button. We would climb all over the dunes to the east of Pensacola beach. I taught all of them how to dive head first from the top of the dune and then land on their hands and roll.

At the Beach
Top: Cathy Second Row: Terry, Bonny (Neighbor), Molly
Bottom: Pinky, Sabra, Dolores, Kenneth (Neighbor)

When we went to the beach or the bluffs on the scenic highway in my car, we would drive with the windows open. I had no horn, and I trained the girls to shout "Beep-Beep" whenever I pressed on my horn button. We would climb all over the dunes to the east Pensacola beach. I taught all of them how to dive head first from the top of the dune and then land on their hands and roll up in a ball going down the dune. We would play hide-and-go-seek, and that never resulted in kissing opportunities. Pinky was always present and always watching. Pinky was everywhere! I now own a high-rise condo overlooking those same dunes and the beach where I courted my bride and her sisters. It is a very sweet memory.

A county fair came to Pensacola and I took our girls out to the fair for some rides and games. One of the games had a lot of stuffed animals for prizes, and after some unsuccessful tries at knocking down a coke bottle, I figured out how the huckster could always do it and we never could. To win a prize, you had to guide a ball on a long string so as to miss the bottle swinging out, but when the ball swung back, it had to knock over the bottle. The laws of physics were against that, since it would miss the bottle by the same distance that you missed it swinging the ball outward. I noticed that the huckster would bring the ball in on an angle barely past the bottle and then spin it off straight out with his thumb, thus the outbound path encompassed the bottle and therefore would swing back and knock down the bottle on the return swing. I took the girls home and set up the ball on a string with the coke bottle directly under the suspension point. The girls practiced all morning and we returned to the fair that afternoon. All of the girls won a stuffed animal, and 5-year-old Cathy had won two, when the huckster finally refused to allow us to continue playing. We had broken the code and it felt good.

Navy Discharge

I was discharged from the US Navy on August 3, 1950. The Korean War had just started and our Secretary of Defense, Louis Johnson, had drastically reduced our armed forces one year earlier. The reserves were being called up again, and people were being drafted. I felt terrible because I knew that I was meant to fly and could fly better than my Navy classmates, and now I was not going to be able to prove myself in a war. The Navy would not pay for me to travel home by car, because it was illegal for a Midshipman to own a car, so I had to go to work for my gas money at the downtown gas station owned by Sherrill Oil Company.

I took a room at a boarding house on Spring Street and made enough money pumping gas to pay my room and board, support my 1940 Ford, and date Sabra. Our courtship continued as before, except I really felt like a "nobody" without a future. Sabra still loved me even though I was working in a lousy job. I finally determined that I was never going to work myself out of this hole if I stayed in Pensacola. It was a town run by a bunch of good old boys, their sons and grandsons, and I could not envision even an adequate future for Sabra and I in Pensacola. Most of all, I wanted to fly ... and I wanted to fly in this war. Since I had been eliminated from flight training, I was not eligible to apply for flight training in the Army or Air Force, so I decided to try Canada; but first I needed to go home to St. Louis and operate out of my own hometown.

Sabra's father was even more convinced now that we should reconsider our engagement and we reluctantly agreed. I left Pensacola with a heavy heart, but I wanted to start on a new future so that Sabra and I could look forward to something more than a mediocre income in Pensacola. I knew she loved me and would wait for me, and she knew that I had to find my way back into the cockpit or an equally challenging career.

Friends

Phillip Joe had been having trouble with his grades at Catholic High School and his father had enrolled him in Gulf Coast Military Academy in Biloxi, Mississippi. On my way to St. Louis I was asked to drop Phillip Joe off in Biloxi. We talked a lot during that period, and I could see that his father was putting a lot of pressure on him to succeed as the only son in the family. His father had to leave school early to help support his family; however, he read law books, studied books about the hotel and insurance businesses, and became a self-educated man who highly valued a good education. He wanted all of his children to receive and value a good education. He had placed Phillip Joe in this military school earlier and his grades had improved. Phillip Joe was depressed to be sent back to Biloxi, and we had a long meaningful talk. I left him in Biloxi and knew that I had a good friend in my future brother-in law. We were both under pressure from his father and it was good for both of us.

I had dated Nancy Lichty quite a bit in high school and we were still good friends. Her father was a Brigadier General in the Missouri National Guard and Nancy agreed to arrange a meeting with him. I told him the entire story of my brief naval career. I told him I knew that I was a good pilot and wanted to fly in this war and asked if he could help me get into the Royal Canadian Air Force. He had heard Nancy talk about my working on the railroad in the North Woods and blubbering seals in Alaska and knew that I had been working in the neighborhood stores since I was eleven. I had met him and talked to him when I was dating Nancy, but I was only one of Nancy's many dates. He and I liked each other and I felt comfortable with him. He promised me nothing, but said he would look into it.

Waiting For The USAF

A few weeks later he called and told me that he had arranged an interview with the local officer in the USAF Recruiting office. I went downtown and he helped me fill out an application to enter pilot training as a USAF Aviation Cadet. I filled out the form honestly, responding yes to the question if I had ever failed to complete flight training before and listed the reason as Honorable Discharge US Navy. I took a battery of mental and physical coordination tests at Chanute Field, Illinois, and two months later I was notified that I had been selected for Aviation Cadet training. I would have to wait about a year for a class, because there were so many other applicants ahead of me. The large draft quotas had caused a lot of eligible young men to apply for pilot training, which appeared to be more attractive than combat infantryman, so I had to wait my turn.

I was extremely happy and called Sabra, and we became engaged again. Her father allowed her to come to St. Louis—with Phillip Joe as her chaperone—to meet my family. My mother, grandmother, aunts and uncles all loved her instantly. (My uncles even wore nice shirts to the dinner table!) My grandmother and Sabra would take long walks in the morning, and my mother would accompany us out in the evening. I was working selling Guardian Service Aluminum cookware, doing home-cooking demonstrations whenever I could talk a friend's mother into sponsoring one of my dinner demonstrations and sales presentation. I worked very little during Sabra's visit and it was a great success. My family loved her very much.

Sabra, Mom, Phil and Doris

All of my friends and classmates had graduated from college by now, and there was a subtle difference in several of them. I had not completed college and was perceived by some to be destined for a less promising future. For the first time in my life I was being considered a dropout, and it made me very determined to work harder and out-produce my entire peer group. My happy-go-lucky attitude, bad temper and undisciplined habits had resulted in my current dilemma and I vowed these traits would be left behind with my carefree youth. I had found my love, and now I was determined to make her proud of me.

It is a defining moment in your life when you suddenly find all of your friends are launched on their career path and you have only two years of college with a job selling pots and pans.

I looked back on all of the opportunities that I had taken for granted. I recalled that I had been evaluated as a near genius but had failed to study hard enough to make the honor roll. I had been a star athlete but failed to practice enough to break records. On the positive side, I had worked hard since I was eleven, had a good work ethic and seemed to be able to obtain interesting jobs where I could satisfy my thirst for adventure. Best of all I was engaged to Mary Sabra Sanchez. I vowed to apply all of my talents to pursue my ultimate ambition to be a fighter pilot and then marry Sabra. I now had the focus that I needed to drag myself out of the hole.

Our courtship continued, but mostly by mail and telephone. I took a job as a traveling salesman, introducing a new product to drugstores in Illinois, Indiana and Ohio. Rid was an iodine powder that cleared up skin blemishes. I would come into to a city after a radio-advertising barrage of one-minute spot commercials that continued for a month, I would offer the druggist three free spots if he would allow me to put up a window display and stock at least one dozen bottles of Rid, which he did not have to pay for until he reordered. I stocked Rid in about half of all the stores that I visited and built window displays in about half of those. After six months, the other salesman working in the district was fired; two months later I was fired, and I believe the company went bankrupt. I guessed that they had to pay for all of the radio ads but failed to get enough reorders to satisfy the outgoing cash flow. My career as a patent-medicine salesman was finished ... but my Air Force career was about to take off.

A Lesson Learned

Sabra came up to St. Louis for one more visit before I went into the Air Force and it was wonderful to see her. We wrote to each other almost every day, but we had a lot more separation ahead of us before we could be married. Aviation Cadets were not allowed to be married. It would take about 13 months to win my wings and be commissioned, and I had already been waiting for a class assignment for 14 months—we had been separated most of that time. I had been occasionally dating two of my former girl friends that I had gone to high school with—Barbara Siegel and Nancy Lichty. Both had graduated from college and had steady boyfriends, whom they eventually married, and both boyfriends lived out of town. I had liked both of them very much and perhaps I was dating them to make certain that I truly wanted Sabra for my wife for life. Sure enough, both turned out to be good friends but not a future wife. Sabra knew that we were just good friends but she still was mad that I was dating them. She informed me that she intended to date occasionally while I was away in the Air Force and she did. I had left myself wide open, and I was miserable thinking about her dating. She broke a few Navy hearts while I was going through Air Force pilot training but never became seriously involved; and I learned my lesson.

USAF Aviation Cadet, Active Duty

I reported for active duty as a USAF Aviation Cadet in January 1952 at Goodfellow AFB, San Angelo, Texas. I only had one anxious moment in the long processing line: A young lady noted that I had previously received US Navy flight training and asked me if that was true. That was probably her only job in the processing line and she probably had never encountered anyone who had. But they apparently had not told her what to do if she did. She was looking around for someone in authority to see what she should do, while the line kept moving; and thank God she was still bewildered when I finished processing and walked out of the door!

I had to go through the entire program again with all of the square meals, harassment and close-order drill, but I was grateful for the opportunity. This time I suppressed my sense of humor and was deadly serious; and I excelled in the academics, military bearing and most of all the flying. I was designated as the Wing Commander for all of the cadets at Goodfellow AFB when our class became the most senior class. When we completed primary training in the T-6 propeller Trainer, I was assigned to Laredo AFB Texas to become a jet pilot, flying T-33 jet trainers. This virtually guaranteed that I would be assigned to be a jet fighter pilot when I graduated and go to war in Korea. I was going to achieve my goal! I was back on top of the world.

Graduation Primary Training
Goodfellow AFB, Texas
Joe Blake, My Mother, Sabra, Neil Leeper

Sabra received permission from her father to come down for my graduation from primary training and was chaperoned by her cousin, Father Henry Thorsen, a brand new Catholic priest. My mother also came, and after graduation we went to Monterrey, Mexico, for a one-week vacation. We had a very good time together and saw all of the sights. There was a Mexican band at our resort, and on the first night they learned that we were from St. Louis. They would play the worst rendition of the Saint Louis Blues that I ever heard every time we entered the dining room! They also had the Mexican dancers take us out on the dance floor during the floor show and dance to the Latin music. On another occasion we were having a cocktail in the bar and a Mexican gentleman farmer became enamored with my mother. Soon he was buying her drinks and she was looking to us to rescue her, but we ignored her. She finally shook him and chastised us for not rescuing her earlier. She was not amused then, but she told everybody that story for the rest of her life. We had a wonderful week in Monterrey; then I reported in to Laredo AFB, and Sabra went back to St. Louis to spend another week with my family before returning to Pensacola. She met my sister Doris and her wonderful husband, Phil Jackson, and they both loved her immediately. They called me to tell me that they enthusiastically approved of my bride-to-be.

We were the first class to be trained at Laredo AFB. The base had just been reactivated after being closed since WWII. The Officers Club and Cadet Club were not opened yet, but we had an excellent chow hall with good food. The old wooden barrack that we lived in was not sealed to the foundation, and the desert winds blew dust into our rooms. We were always receiving demerits from our tactical officer for dusty hat, shoes, and footlocker every time a wind came up and would have to march the demerits off by walking tours on the parade grounds. Laredo AFB was an austere base with few amenities in the middle of a very big desert. The town of Laredo was a typical border town with very little industry and very few nice places to eat or drink. Nuevo Laredo was just across the Mexican border and catered to tourists who came down to shop, eat Mexican food and partake of various other activities.

Secret Plans

Sabra and I had decided to be married during Christmas leave, because we knew I was going to be assigned to Korea after I received my wings and completed fighter-pilot training in my assigned combat aircraft. It was forbidden for cadets to be married until they were commissioned and received their wings. I was scheduled to complete training and be commissioned on February 2, 1953, so we planned for our secret wedding on December 22, 1952—we would only be "illegal" for five or six weeks. We had been engaged for 27 months and separated most of that time. We were both eager to be married and knew that we were facing another year of separation while I completed my combat tour. This would provide us with about 22 weeks of married life together until I won my wings and became a fully qualified combat pilot. An extra six weeks of married life with Sabra was worth the risk of discovery.

The Infamous Episode

Six of our instructor pilots took their wives out on the town shortly after the base reopened. There were no officers clubs at that time, and the restaurants in Laredo were very austere, so they drove across the Mexican border looking for some flashier restaurants. Some of the larger and better known night clubs were in the red-light district, so they took their three-car convoy into the district and went into one of the larger nightclubs where there was entertainment. They took a large table for 12 and asked for menus. The Mexican Police were there in a very short time and arrested the six wives after a heated discussion with the husbands. There was a language problem and they all felt that it was a communication problem until the police loaded their wives into a paddy wagon and hauled them off to the Nuevo Laredo jail. The officers finally had an English-speaking officer explain that the red-light district was off limits to any females who were not employees or licensed prostitutes. The officers retreated across the border and called their commander; and the base commander and a lawyer started negotiations with the police and, finally, the Nuevo Laredo mayor. They reached a solution the next morning: The young officers all paid a fine and purchased prostitutes' licenses for their wives. Thus we had six wives in the officers wives club licensed to practice prostitution in the red-light district in Nuevo Laredo, Mexico! They had spent a miserable night in a Mexican jail without dinner or breakfast but had a memento to hang on the wall for the rest of their lives. The six couples were the victims of many jokes for several months and, before long, Laredo had a new officers club. One of the officers was an instructor in our squadron and he and his wife were still being kidded about that infamous episode when we left Laredo.

One Last Crisis

Our courtship ended when I departed Laredo AFB for Christmas leave, enroute to Pensacola, Florida, to finally marry the Lady of my dreams. My last crisis occurred two days before our leave was scheduled to start. All Christmas leaves were suddenly delayed until our class flew enough hours to catch up to our schedule. They anticipated a two- or three-day delay, which also would delay our wedding. I immediately went to our flight surgeon and told him about our illegal wedding plans. He promptly grounded me for a badly sprained ankle, shaved my leg and taped my ankle up to my knee as a wedding present. The Squadron Commander signed my leave papers, and I departed on schedule for Corpus Christi, Texas, where Tex and Annie Atkinson (Sabra's older sister) were stationed. Tex was now a full Lieutenant instructing new USN pilots to qualify as combat pilots in the F-6F fighter. They would drive me to Pensacola and be in our wedding party.

I arrived in the late afternoon, and Tex was not home yet. When he arrived, he brought one of his student pilots with him— one who had dated Sabra and fallen madly in love with her. Danny Parker was a short man with a very good

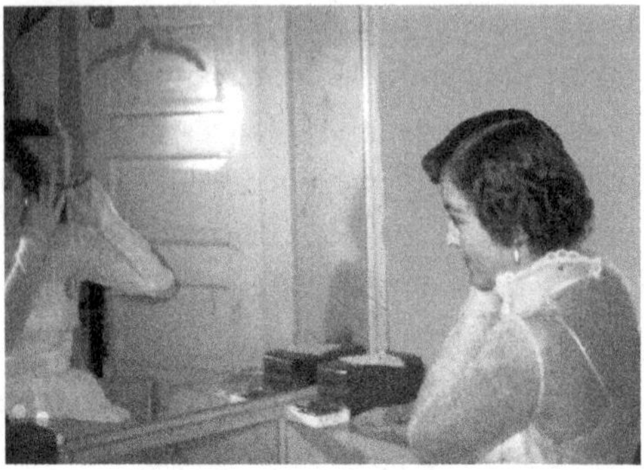

sense of humor, and he proceeded to tell me how lucky I was to be marrying Sabra. Tex kept pouring drinks, and I was sitting on the end of my chair, anxious to start driving to Pensacola. The wedding was scheduled to take place in two and one half days, and we had a 12-hour drive ahead of us. It turned out that Danny Parker was also going to ride with us. By the time we ate dinner and they packed the car, it was about midnight when we departed for Pensacola.

Tex drove to the outskirts of Corpus Christi and pulled over to the side of the road to take a nap. Danny and Annie were already asleep, so I decided I would drive. I had been awake since 4 a.m. and was dead tired; but this was an emergency, so I took the wheel and started for Pensacola. Sometime in the middle of the night the engine began quitting and I became more alert. I was out of gas in the middle of Louisiana. I was so sleepy that I had paid no attention to the gas gage and we were completely empty! I got out of the car and got an empty gas container from the trunk and started walking and hitchhiking towards Pensacola. I could not remember passing an open gas station recently, and we had been in the remote part of the state for a long time. A trucker finally picked me up and drove for about 20 miles before we found an open gas station. I bought the fuel and started hitchhiking back towards the car. Another trucker picked me up and told me that it was 40 miles between gas stations on this section of remote Louisiana, so I had picked the worst possible spot to run out of gas. This delayed us for nearly two hours but we were off again. At some point in time, Tex or Danny woke up and took over the driving and I fell sound asleep for the rest of the trip.

We arrived in Pensacola before noon, and as soon as I saw Sabra everything became happy again. I learned that Sabra's Grandma Sanchez had died earlier in the week, and her father's brother, Lars, had tried to cancel our wedding. Sabra's father, however, handled him. Thus, the week before our wedding was very stressful for both of us, but we knew we were only two days from our long-awaited marriage, and all of the stress melted away. Twenty-seven months and two weeks after I proposed, we were finally going to be married—and I was well on my way to achieving my goal to become a fighter pilot.

CHAPTER 7

Our Wedding

Tex, Annie and I arrived in Pensacola the day before the wedding. My mother, my sister Doris, and her husband Phil Jackson were already in Pensacola when we arrived. Much to my regret, my wonderful grandmother could not make the trip. Doris, Annie Atkinson, and Dolores Sanchez were the bridesmaids, and Phil Jackson was my best man, with Tex Atkinson and Phillip Joe Sanchez as ushers. I did not know that Grandma Sanchez had died a week earlier until we arrived in Pensacola, and that reduced the festivities to some extent. Our families got along fine with each other; and, as usual, Sabra's father was a most gracious host. At long last we were to be married. The wedding date fell on my birthday, 22 December 1952, just three days before Christmas. It was the best birthday present I ever received. Since I was not Catholic, we were married in front of the altar at St. Stephen's Church by Father Royer. We had a rehearsal that night at St. Stephen's and then had a nice dinner with the two families. All of the little sisters were in attendance and looked beautiful. All of the children were very happy that we were being married. They were not losing their little mother but gaining another big brother.

Just Married

The wedding went very fast. I was a nervous wreck. When Phil Jackson, Phillip Joe and I came out from the side after the bride was brought to the altar by her father, we were traveling very fast and were very close together. My sister commented later that it looked like the Marx brothers coming on stage. It was a definite change of pace following the beautiful bride, escorted slowly down the aisle by her father. Sabra's father gave me her hand (much to my relief) and the wedding proceeded. I envisioned that we would have to take all of the little sisters on our honeymoon just as we had on most of our dates.

The reception was in the Bay Shore Room of the Bayou Vista Hotel, her father's hotel, one block away from their home. I cannot recall any details, but everyone had a very good time and our families mixed very well. I remember that Phil Jackson had to borrow shoes from Phillip Joe because he had forgotten to bring his size-14 dress shoes. Phillip Joe wore a size 12 and Phil could not wait to get out of those shoes! He went shoeless the rest of the night. I also remember Phil Jackson and I playing a duet on his Tonettes, which is remarkable, since I cannot play any musical instrument.

Sabra's father had rented us a room in the Battle House Hotel in Mobile, Alabama, and told us that nobody knew where we were going. He had managed that hotel briefly some time ago and had arranged for a nice room. Dolores and her future husband, Walker Wilkins, had hidden our car to prevent a sabotage job and drove us to where they had hidden it. Sabra had recently made a down payment on a new 1952 Plymouth two-door sedan, and we now had our first car.

Walker and Dolores followed us out of town, and soon we smelled burning rubber. I stopped and found a lot of foul-smelling smoke coming from our exhaust pipe. Someone had put a rubber shoe in it. The culprit? Walter! After I pulled the shoe out, we proceeded to Mobile, checked into the hotel and began our wonderful marriage. I remember wondering out loud whether or not to remove my wristwatch before going to bed. Sabra, with a sparkle in her eye, reminded me that it was shockproof.

We were finally husband and wife and had declared our love for one another in front of God and all of our family and friends until "death do us part." It was a very beautiful and solemn ceremony and we both took it very seriously. I vowed to make our marriage as good or better than that of my grandparents, which was the only successful marriage in my family. Sabra and I loved each other very much and knew that ours would last. Our total trust in our love and mutual commitment has been an important part of our very happy marriage.

Very early the next morning we were awakened by the telephone. I could not believe anyone would call us, because only Sabra's father knew where we were. It was Sabra's Aunt Wissie, the nun, and she wanted us to drive her and her friends around Mobile to shop for a few things. They had finished breakfast and said their prayers and were ready to go! I still could not believe it, but Sabra and I leapt out of bed, dressed and drove to the convent to pick up Aunt Wissie (Sister Terisita) and two of her nun friends. They also wanted us to sing while we drove them around, just like we did when we would drive her mother and Aunt Rachael to Mobile to visit Aunt Wissie and her friends while we were courting. Well, at least we did not have the little sisters on our honeymoon! I believe Sabra's mom and dad put Aunt Wissie up to it, because I never saw her eyes sparkle like they did that day. I bet those nuns told that story the rest of their lives.

Aunt Wissie was happy that we were married, even though I wasn't Catholic. She gave me a silver Saint Christopher Medal on a silver chain that day. I wore it on all of my combat missions throughout my career. That was more than 50 years ago, and I still wear it around my neck every day and rub it and whisper Aunt Wissies' communion blessing when the priest blesses the Bread and the Wine before communion.

We stayed in Mobile three days, and then went back to Pensacola to join the family festivities until my sister, her husband, and our mother left to drive to St. Louis. We all celebrated Christmas at Sabra's family home and had a wonderful time, but I cannot remember anything except being very happily married to Sabra. After New Year's Day, we drove off to our first duty station, Laredo AFB, Texas, to complete my jet pilot training and be commissioned as a second lieutenant, USAF.

There was one small problem: We were not allowed to be married until I won my wings and was commissioned. The next eight weeks were going to have to be very austere on my $105 monthly salary and with a hidden wife! Aviation Cadet and Mrs. Robert W. Kennedy were on their way to the first of many duty stations.

My Mom, Me, Sabra, my sister Doris, and her husband Phil

Sabra's Mother, Me, Sabra and Sabra's Father

CHAPTER 8

Wings, Sabre Jet Fighter Pilot, and Las Vegas

Before leaving Laredo, I had arranged to rent a room in the Kennedy Motel (no relation) at an affordable $80 monthly rate. It had a double bed, a wooden chair, a small dresser and a very small ice box. You placed a block of ice in the top every three days and it kept a small amount of food cool. There was an orange tree outside our door, so Sabra had her source of fruit in our little Garden of Eden. It may have been our Garden of Eden, but one month later Annie drove their mother over from Corpus Christie to visit Sabra for the day. When their mother saw our motel and the room, she broke out crying. She finally calmed down, and we restored her good humor before they returned to Corpus Christie.

There was one other married cadet from my class with his new bride staying in the motel with us. Several other cadets in my class had planned their weddings over the Christmas holidays to have some time together before going to Korea. It was an austere beginning.

We were still Aviation Cadets subject to early morning roll call and inspection, the daily flying schedule, and meals in the Cadet mess. Both of us would spend nights with our wives in the motel, and then put on our cadet uniform with our fatigue jacket and cap over our uniform. We would drive through the base gate about 4:30 a.m. with the early morning aircraft maintenance shift, park the car in a lot near our barracks, leave our fatigues in the trunk and walk to our barracks. We would go to our cubicle and remake our bed, which we had stuffed the evening before and await the morning muster and inspection. Our roommates obviously knew our secret and assisted in the deception. I became very skilled at stuffing my bunk, which I would do right after the evening meal. My blue flight jacket had a fur-like collar, and I would curl that up on the pillow so it appeared to be my crew cut. Nobody ever commented on the blue hair. We would return to our new brides early in the evening and bring them food smuggled out of our chow hall. We were Aviation Cadets from 0430 until 1800 six days a week and loving husbands the rest of the week. We had a few nights of night flying.

One day my flying instructor, Captain Wise, saw Sabra and me in town. He smiled, waved and went off trying not to laugh. He never mentioned it, and on graduation day he came over and congratulated me and asked to be introduced to my wife. He told me never to volunteer for undercover work, because he figured me out within the first week. He said I glowed in the dark!

We were very happy at Laredo. The lack of things to buy there never bothered us, because we had no money to buy anything. On Sundays we would walk across the Mexican border to Nuevo Laredo and watch the people. Sabra still had her maiden name on her driver's license and would show it when we crossed the border. The Mexican border guards always smiled at her and looked her over very well, then would check her license and pronounce her name very slowly, Maria Sabra Sanchez and resentfully check me (Gringo Bob) silently.

We received our orders in mid February, and only six of our class of fifty would be sent to Korea. Apparently there were too many fighter pilots and our pipeline had finally filled the requirements. I had orders to attend fighter pilot training in the F-86 at Nellis AFB, Nevada, and then to the 51st Fighter Interceptor Wing in Korea. The F-86 Sabre was the first and only supersonic fighter in the world; Nellis was the best fighter pilot training base in the world; and the 51st FIW was one of the two Wings in the USAF engaging in air-to-air combat against the North Korean Mig-15 fighters. All Mig aces had achieved their kills in those two wings.

The top six cadets in our class were assigned to the F-86 in Korea, most to the F-86 at stateside bases and about ten to F-94 or F-89 all-weather training, with assignments to the Air Defense Command. I was in hog heaven—all of my objectives, which I'd established when I left the USN, were exceeded and within my grasp; and, I was married to the most beautiful soul in the world. I envisioned returning home from Korea with 100 missions after only six months in Korea as a Mig ace and a spot promotion to first lieutenant. Captain Wise, my flight instructor, was also very pleased and told me he knew I would make it through the rigorous course at Nellis AFB. Most of my classmates were disappointed that they were not going to Korea. We had all been trained to be fighter pilots in Korea—"Every man a Tiger"—and now only a few of us would be given the opportunity. I had many opportunities to trade my combat assignment for an F-86 assignment at a plush stateside base, but I needed to experience the combat to see if I had "the right stuff."

The biggest lesson I had learned that would help me the rest of my life was that I had set firm goals after my disaster with the US Navy and pursued them, focusing all of my activities toward achieving those goals. The accomplishment of those goals was very rewarding. Sabra and I became goal-setters and planners from that point forward, which significantly enhanced the quality of our life. Sabra made a budget and on the first day of the month would set aside the savings required to achieve our goals. Then we would live within our budget. It worked with everything except having children. We both wanted many children, and that turned out to be our biggest challenge.

Onward to Nellis AFB, Las Vegas, Nevada

Sabra and I drove home to St. Louis for a 10-day leave to visit my family and all of my friends. Few of my friends could understand my excitement and pleasure in going to fight a war in Korea. All had served a tour in the service and Tex Laurence had seen a lot of combat in Europe. All of them were proud that I was going to fulfill my lifelong dream to be a fighter pilot. One of my friends, Eldon Brewer, had served in the US Navy, but not long enough, and he had been drafted and was also going to Korea. He had completed college but was not interested in becoming an officer. He preferred doing his minimum time and then getting on with a civilian career. We had a wonderful time in St. Louis. All of my friends loved Sabra. After ten days we were off to Nellis AFB in Las Vegas, Nevada.

We found a small triplex just three blocks from the fabulous Las Vegas strip. I was now a 2/Lt making $200 per month plus $100 flight pay, and we were very rich. We settled into our triplex and I reported for duty. I found that I had been selected to be part of an experimental group under the supervision of Captain Boots Blesse, a double Mig ace. Three Lt/Colonels and three 2/Lts on orders to Korea had been selected for this project, based on our flying scores, and assigned to the 3525th Fighter Training Squadron. We attended ground school with the other F-86 student pilots, but all of our flying was to be in the 3525th, which was the squadron dedicated to training F-86 Instructor Pilots. Many of our instructor pilots were Mig aces recently returned from Korea, and we had more Mig aces in our squadron than any other squadron in the USAF. Captain Blesse's objective was to send three leader/wingman teams to Korea who were trained specifically to fly air-to-air combat. Our training would consist only of air-to air gunnery and air-to air combat against the F-86 Instructor pilot Mig aces in our squadron. I had hit the jackpot and far exceeded any objective that I could have envisioned. I was going to be checked out in the only supersonic fighter in the world and instructed in the art of aerial combat by the best of the best.

I finished the F-86 ground school with no problem and received a cockpit checkout from my instructor, then started the engine. My instructor and I taxied out to the runway, and I took off with my instructor flying my wing. There were no Trainer (two-seat) F-86 aircraft, so the first and all subsequent flights were solo; however, we usually flew in two-ship or four-ship formations as a team. My instructor had me fly a few acrobatic maneuvers, do a few stalls, and then he put me in trail formation to follow him through some maximum-performance maneuvers. I hung in tightly and we returned to the runway. I made a few touch-and-go landings and was checked out in the F-86. I was elated! I was an F-86 pilot, and in the next few months they were going to make me into a highly skilled fighter pilot. From there I would go to Korea and fly air-to-air combat against North Korean and Russian Mig-15 interceptors.

Air-to-air combat is executed by flights of four aircraft in a formation most suitable to the tactical situation, and each flight has a designated flight leader. The flight fights as two 2-ship elements with each element consisting of a leader who is the shooter and a wingman who flies a loose wing on the leader and searches the rear hemisphere for enemy threats to his element leader or other friendly aircraft. Our only weapon was six 50-caliber machine guns, which fired straight ahead, fixed to the aircraft. Thus, the object of each leader was to maneuver his aircraft to point his guns where his bullets would impact on the enemy aircraft, preferably by being directly on his tail. This usually involved very violent maneuvers and high G loads. The wingman must stay with his leader while searching the rear hemisphere, often while "graying out" from the high G load. At 4 to 5G loads the blood drains from your head and your vision goes from clear to gray to blackout. We all wore G suits, which automatically inflated as the G load increased to squeeze the abdomen, upper legs and lower legs, reducing the flow of blood from the head. Otherwise, we would gray out at about 3 to 4 Gs.

The training concentrated on 4-ship and 2-ship formation and tactics, actual air-to-air combat using gun cameras, one aircraft versus one aircraft, two on two, and four on four, and many gunnery missions firing the 50-caliber machine guns at a target towed behind another jet aircraft. At the end of the gunnery flights all flights would fly to the green spot in the desert and engage in random attacks against other flights in the area, often resulting in three to four flights of four, fighting violently to obtain the favorable tail position. It was very fast, crowded and confusing, and the adrenalin flowed freely. This is where the lesser wingmen lost their leaders or failed to cover their leaders. It was best described as a "gaggle," much like actual combat. It was our most valuable training.

The recipe to make a fighter pilot can be oversimplified: 1) Know your weapons and practice often; 2) Learn and continually practice your fighter tactics and four-ship teamwork (actually two two-ship shooter-wingman teams) required to maximize firing opportunities; 3) Spot the enemy before he spots you; and, 4) Be aggressive.

The first few flights were solo flights to familiarize myself with the aircraft and the local area, with a few two-ship formation flights with one of our instructors. I had the thrill of flying down in the Grand Canyon and could not wait to go supersonic. One of our instructors was Captain Ivan Kinchloe, a recently returned Mig ace, and he promised me we would go supersonic on my second two-ship flight. He briefed me on the training mission and then told me that when we returned to the base we would maintain our altitude above 45,000 feet, push the throttle full forward, and when we reached max mach for level flight we would roll over on our back and pull through until we were in a dive straight down. I was to fly loose formation and watch my mach needle when the aircraft began vibrating. When I saw the mach needle go through mach 1.0,1 was to press the microphone button and start screaming at the top of my voice and scream until we hit mach 1.04, which was the max for the F-86. He implied that we would not be able to hear my voice because we were exceeding the speed of sound. Well, I did as directed and learned it is very hard to scream into an oxygen mask at very high altitudes, but I emitted an acceptable scream. I also learned that I heard my scream very clearly on the radio and so did everyone else on our tactical radio channel within a 230-mile radius. I knew that I had been had when an unknown voice on our radio channel stated, "Ivan just suckered another new guy." I was one of several 2/Lts dumb enough to experience Captain Kinchloe's sense of humor. A few years later Captain Kinchloe was test-flying an F-104 out at Edwards AFB, lost his engine on takeoff and died after an unsuccessful ejection. The USAF named an AFB in Michigan after him. It was a great honor for me to fly with this truly great fighter pilot.

Many of our training missions were in a flight of four F-86s that would make several live firing passes on a long banner towed behind a T-33 aircraft. We would sit on a perch about 1000 feet above the towed banner, offset about one half-mile to the side and slightly behind the banner. We would take turns diving down, turning first left and then reversing into a right-hand pursuit curve to approach the target from above to fire live rounds from our 50-caliber guns. We would keep making passes until all four of us had fired our 100 rounds. I really learned the gunnery pursuit curve well from our instructors, so that I could fly and time the maneuver to end up sliding my gunsight pipper down the gunnery target when I was within the 1000-foot range and firing my 50-caliber machine guns while the pipper was on the target. My scores were consistently higher than all of the other students and often better than the instructors. Captain Blesse was a very dedicated and effective instructor. He would spend hours telling us all of the technical details on the six 50-caliber machine guns and the manual and radar gunsight modes and describing them in a meaningful operational context. He was a double ace with 12 Migs destroyed in air-to-air combat and had just recently returned from a very successful 100-mission tour in the F-86 in Korea. His post-flight debriefs and many air-to-air live gunnery missions against towed banner targets provided all of us with the basic gunnery skills to satisfy the first ingredient in the recipe. He had a very successful 35-year Air Force career, contributing much to tactical aviation, and retired as a Major General.

The most important ingredient in the recipe is how to maneuver your four-ship team so that your two-shooter wingman teams end up on the tails of your opponents within the range of your six 50-caliber guns. I was successful at gunnery because I had mastered the pursuit curve so that I was in perfect firing position against my slow-moving, non-maneuvering banner target. The real trick is to out-fly or out-think your opponents so that you end up in an acceptable position within the range of your guns as a result of superior fighter tactics. Unlike the slow-moving, non-maneuvering banner, our enemy fighters would be at max velocity and maneuvering at max turn rates with max G loads. We spent most of our missions flying air-to-air combat against four-ship formations with Mig aces in the lead-shooter positions.

I flew as the wingman responsible for checking our rear hemisphere to warn my leader if we were under attack and most important to stay in formation with my leader regardless of his maneuvers. Most of our missions were in four-ship formations fighting other four-ship formations. As soon as maneuvering commenced, it would become four two-ship elements, each with a shooter and a wingman. Wingmen who could not stay with their shooter would normally cause both to be shot down by the two-ship enemy elements that stayed together. As a wingman, I never lost my leader and occasionally was allowed to fly one of the shooter positions. Toward the end of our training, I flew as the element lead and occasionally as the leader of the flight of four.

The shoot-downs could be verified by scoring the aerial gunnery films, which would photograph the view ahead of the guns whenever the gun trigger was activated. The maneuvers involved full throttle maximum G turns, trying to keep track of 8 to 20 other aircraft all engaged in trying to obtain a firing position on each "enemy" aircraft. We would fly 45- to 55-minute missions under very high G loads and would be too dizzy to walk a straight line when we got out of the cockpit. It was max performance on the aircraft and the pilot for the full mission. It turned out that the F-86 enjoyed a 10 to 1 kill ratio against the Mig during the Korean War, in large part due to the superior training of our pilots. The USAF doctrine is to fight the air war over enemy territory so we always had to travel to and from their territory from our bases. They flew Mig-15 interceptors designed to defend their territory from attack by enemy bombers. Their combat missions were all within their local areas, generally lasting less than an hour. We flew fighters and carried enough fuel to penetrate deep into enemy territory, and our missions were about 2.5 hours. Thus we were always at a weight disadvantage even after we jettisoned our external fuel tanks and their interceptors could always climb faster and turn inside of our fighters placing us at a disadvantage.

We could always outdive the Mig. The Mig became difficult to control at high speed and lower altitudes. The Mig-15 was designed to kill enemy bombers and was armed with very slow-firing cannons, which could bring down bomber aircraft with a few hits and fighter aircraft with a single hit. We were armed with six 50-caliber rapid-fire machine guns, which put out a lethal stream of bullets more likely to hit small maneuvering aircraft but requiring more hits to effect a kill. Thus our guns were more effective than their cannons in a wild maneuvering air battle because of our much greater rate of fire. Our best advantage stemmed from our superior training, aggressive behavior and freedom of operation. The Mig-15 pilots were under control of their Ground Radar controllers who vectored them to their targets, and cleared them to fire and engage. Their pilots were usually controlled from the ground and our pilots played it by ear as the air battles progressed and were far more aggressive. When I completed my training at Nellis, I knew I was destined to be a Mig ace with the confidence possessed only by brand new 2nd/Lts.

Our accident rate at Nellis AFB was very high and became the subject of unfavorable press in early 1953. We only lost one of our classmates, 2/Lt Hugh Grace, who crashed an F-86 in the landing pattern while we were at Nellis, we also lost one cadet in basic training, Cadet Dan Nash, who crashed a T-28 in the landing pattern at Goodfellow AFB, Texas, and two cadets and an instructor pilot in a mid-air collision between two T-33 jets practicing air combat maneuvers near Laredo AFB. I still remember them in my prayers before mass.

Sabra and I had a ball while we were at Nellis. We were the only married couple in our class. All of our bachelor classmates lived in the barracks on Nellis AFB, while we lived two blocks from the famous Las Vegas strip. We would fly from sunrise to about 3 p.m. because the aircraft became too hot to touch in the afternoon desert sun. The bachelors would go back to the barracks and go right to bed, arising about 9 p.m., then heading downtown to chase the chorus girls. Before long they all had chorus girl friends. Most of them were young college girls and all of them were most attractive. The bachelors would come to our house about the time Sabra and I were going to bed and stay until midnight, when many of the big shows were over and the chorus girls were available. On weekends we would accompany our classmates on a picnic with their dates and had some wonderful parties. Ray Swift, Ray Cheek, and Wes Shipstead all dated lovely intelligent chorus girls and we had a good time.

After 16 weeks at Nellis AFB we were qualified as F-86 Fighter Pilots and departed for our assigned bases. Wes Shipstead, Ray Swift and I were all assigned to Suwon AB in Korea and Wes Shipstead and I were assigned to the same fighter Squadron—the 39th Fighter Interceptor Squadron, 51st Fighter Interceptor Wing. We all graduated as F-86 fighter pilots, and six of us were assigned to Korea.

Sabra and I drove back to my home in St. Louis to visit my mother, grandmother and Aunt Gladys, and then I drove Sabra to her home in Pensacola to live with her parents until I finished my 100 missions. I left her with her family and flew to Camp Pendleton, California, to await transportation to my Fighter Wing in Korea. Sabra managed her father's Bayou Vista Hotel and restaurant during the time that I was in Korea. I started missing her as soon as I left Pensacola.

Sabra, Momo, Gladys, Mom at my home.

CHAPTER 9

Korea

Wes Shipstead and Ray Swift were already at Camp Pendelton, California, when I checked in; we learned we would be there a week or so until they could fly us to Korea. We spent the week playing poker, drinking and reading, and finally took off on a crowded Constellation transport aircraft to Japan. The military had not yet built any jet transport aircraft. I believe we had two refueling stops and the last one was Clark Air Force Base in the Philippines, so it was a long flight to Tachikawa, Japan. We then transferred to an enormous four-engine C-124 transport (We called it the crowd killer!) and flew to Seoul, Korea. The whole trip took about 22 hours. We were weary but excited to finally be in the war zone.

Ray Swift was picked up at the terminal by a jeep for transport to his squadron, and Wes and I had to call our squadron for pickup. We were stationed at Suwon Air Base—about 30 miles south of Seoul—and the jeep was going to take one to two hours to traverse the road before we would be picked up. Wes and I decided to take a walk into Seoul to see how much damage had occurred from the invasion by North Korea and the retaking of Seoul by our UN troops, which were mostly American.

There was a lot of damage, but many of the military facilities had been repaired. The Korean civilian housing outside the base was in very bad shape and was patched up, but it was far from repaired. The streets were muddy and the Korean civilians looked tired and thin; however, they smiled at us and seemed friendly. It was growing dark, so we started back to the base. It was then I spotted an older woman doubled over, squatting on the ground in a narrow alleyway. She seemed to be in pain, but nobody was helping her. I left Wes and went to help her get up, to see if I could determine how badly she was injured. I was surprised when she vigorously resisted my assistance. ZOUNDS! It suddenly dawned on me: She was having a bowel movement! Only a brand new Second Lieutenant in Korea would create this kind of embarrassment. We exited stage left and returned to the terminal for pickup.

The jeep ride back to Suwon was an adventure. The road was occasionally surfaced, but there were many stretches of mud over a very bumpy surface. I now understood why it took our driver so long to drive 30 miles. Finally we arrived at the gate, which was guarded by two South Korean Army troops. They checked our ID cards, and we drove on to Suwon Air Base, code name K-13, our new home.

The jeep dropped us off at our flight barracks. We were both assigned to the 39th Fighter Interceptor Squadron. I was in D Flight and Wes was in B Flight, so we lived in different barracks. The "barracks" were long rectangular structures made of uninsulated corrugated steel. There was a door at each end with a clear aisle between the living areas and two round iron stoves in the aisles, which burned wood or coal as our winter heating systems. There were about seven living areas against each of the two long walls. Each living area was equipped with a single fold-up Army cot, two khaki Army blankets, two sheets, a pillow, a wooden tip tank shipping container made into a closet, another smaller wooden container as a desktop/dresser and a desk lamp. Home at last.

D Flight had just had significant changes. Captain Joe McConnell, the leading Mig ace of the Korean War (15 Migs) rotated home to George AFB California after his very successful tour. He died in an F-86H crash a few years later. Captain Harold Fisher (13 Migs) had been shot down chasing a Mig over the Mig base in China. He was held as a political prisoner along with nine other USAF aircrew members who had been shot down over China.

Our UN POWs were returned to South Korea after the war ended in 1953. The Chinese did not return Captain Fisher and the other nine USAF political prisoners until 1956. Captain McConnell had been the D Flight commander, and he was replaced by Major Jack Bolt USMC, who had six Migs to his credit and was also a World War II ace. (An ace is defined as five or more air-to-air kills.) We also were blessed with an experienced RAF pilot, Flight Leader Roy

Home

French. The other ten members of D Flight had all just graduated from the same pilot training program that I had just completed.

Second Lieutenants Wendell Wood and Bob Groszer received their wings in the class six weeks ahead of me and were just beginning to fly combat missions. First Lieutenants Don Bordel and Bob King were 12 weeks ahead of me, had more than 10 combat missions, and therefore had already received their spot promotions to First Lieutenant. (It takes 18 months to make First Lieutenant; however, a combat commander can reward a Second Lieutenant with a spot promotion after 10 combat missions.) First Lieutenants Dean Abbott, Killer Kilbride and Cooky Cook were 18 weeks ahead of me and had about half of their 100-mission tour completed. (A tour was considered complete after 100 completed missions or one year in Korea.) Dean Abbott flew many of his missions as Captain McConnell's wingman, which was a great honor. First Lieutenants Fritz Meyers and Jerry Carlile were 24 weeks ahead of me and had more than half of their missions completed. I met all 11 of my D Flight comrades that night and was very proud to be a member of their team.

Lieutenant Colonel William Beale had been in the special group of six at Nellis AFB training to be shooter/wingman team members. When we both received orders to the 51st Fighter Interceptor Wing (FIW), I assumed that we would carry out Captain Blesse's objective of delivering a well-trained shooter wingman team to the 51st FIW. L/C Beale was already in the wing as deputy wing operations officer and flying combat missions. He had come directly to Korea from Nellis, and I had taken 30 days leave before I arrived. I hoped he would use me as his wingman, but he never initiated the effort and I was not about to suggest it. I'd had the best gunnery scores of the six at Nellis AFB, and he had the worst. I'm sure he did not want to discuss this experience, and I never revealed it to anyone until now. About two months after I arrived, our Squadron Commander L/C George I. Ruddell (6 Migs) returned to the States and L/C William Beale became our new Squadron Commander.

L/C Ruddell was an excellent commander and a very straight arrow. The rules of engagement forbade us to engage the enemy outside the boundaries of Korea. Many Migs—some would say most of the Migs—were shot down across the border from North Korea in Manchuria (China); several were shot down in their traffic pattern in Manchuria. The American air superiority was so overwhelming that all of the enemy fighters based in North Korea were moved into Manchuria sometime after 1951 because we controlled the air over North Korea. We enjoyed a 10 to 1 kill ratio over the Migs. Many of our F-86 pilots would cross into Mancuria to search for the Migs because they were few and far between over North Korea. L/C Ruddell vigorously supported the rules of engagement and would not cross the border. He shot down all of his Migs over North Korea and was highly respected. I would have gone wherever my flight leader went, and there was a greater chance of finding Migs across the border. Unfortunately, I never had the opportunity.

When I arrived, it was very clear from the peace talks that the war would end before I could fly even the 10 missions required to receive a spot promotion to First Lieutenant. A further complication was that a policy was established that if you had 50 missions when the war ended, that was one-half of your one-year combat tour; so, you would go home when you completed six months in Korea. Therefore, everybody was struggling to maximize their combat mission count to minimize peacetime occupation duty in South Korea. They had more pilots than required to fill the flying schedule, so they cut off the new pilots with no missions and evenly distributed the remaining combat missions to those pilots who already had combat missions. It was a fair system, but it meant that I was destined to occupy Korea for one year as a Second Lieutenant and go home without firing a shot—a far cry from my Walter Mitty fantasies. I flew enough to receive my flight pay but only over South Korea. The assistant operations officer took me across the front lines in an unarmed T-33 the day before the war ended, and that was my only claim to fame. I doubt if anyone even shot at us.

I was assigned a bunch of additional duties as a new lieutenant, and was lucky enough to be designated the Squadron Armament Officer and Gunnery Officer. There were about 50 enlisted armorers headed by a very capable Master Sergeant, and he taught me all I needed to know. Our only armaments were the six 50-caliber machine guns. We had to maintain these and periodically bore-sight the machine guns so they all were pointed along the optimum vector that increased their lethality. We also maintained and replaced the cartridges that jettisoned our ejection seat and canopy and were responsible for launching and recovering the aerial tow banners that we used to practice gunnery.

As the Squadron Gunnery Officer, my duties were to learn how to assess gun camera film, maintain the film library and teach others how to assess film. The film library had all of the squadron gun camera film, and I reviewed all of our squadron Mig kills that were recorded on film. Most of the kills were inside of 1,000-feet gun range and visually depicted the hits on the Mig, the resulting debris streaming from the Mig, and the danger of flying through the debris, particularly if the Mig was not maneuvering. One showed the Mig pilot ejecting from his aircraft. Several shots showed a few hits at long range (probably using radar ranging), which slowed the enemy down, followed by a kill at very short range. The high "G" maneuver kills/hits/misses are inferred because of the movements of the gunsight reticle and the background scene on the forward canopy combining glass, which reveal the maneuvers and the "G" load. Viewing the combat gunnery film just whetted my appetite. I learned a lot about our guns and air-to-air combat gunnery from both of my additional duty jobs.

I also learned that I was sorely lacking in experience. When I took over the armament officer job, I had to sign for all of the inventory. I knew that I could be held financially responsible for the inventory, and it was totally inaccurate on many important items. Our wartime supply system was apparently not responsive, and a lot of trading occurred between many units to satisfy urgent needs. We always had two Marine pilots attached to our squadron and traded a lot of goodies with their nearby USMC squadrons.

We routinely carried fuel tanks under our wings because we had to fly long distances to our battlefield up near the Manchurian border. Whenever Migs were spotted, all of the F-86 aircraft would jettison their under-wing fuel tanks. North Korea was littered with jettisoned fuel tanks. (We normally flew above 45,000 feet, and the tanks would come falling down to the ground, reflecting sunlight as they tumbled. This probably resulted in the false North Korean charges that we were dropping chemical and/or biological weapons.) We failed to adequately account for all of the missing tanks and all of the trading deals with other squadrons. Shortly after the war ended, a supply team from "on high" came down for an inspection. All of us who had signed for the inventory we should have had were in very deep yogurt. We had lost our wartime wing commander, Colonel Gabriel Gabreski, who was a WWII and Korean War ace (5 Migs), and he had been replaced by Colonel Benjamin O. Davis, who had commanded the first all-black fighter squadron that distinguished themselves during WWII. I am not sure how Colonel Davis saved us, but we all got off without letters of reprimand, and we were all forever grateful. Most responsible officers noted the shortages when they signed for the account ... but not I. It was a very serious mistake and I learned another lesson. I served with Colonel Davis again in Europe (1958-1962) when he was a Brigadier General and in Thailand (1967-1968) when he was a Lieutenant General. Many years after he retired, President Clinton promoted him to full General. He was a true gentleman and a wonderful leader.

The great majority of my time was spent as a line pilot in our fighter squadron. I was integrated into my flying duties as soon as the war ended. My armament troops obtained gunnery-banner targets and towing cables, and rigged up a two-cable hookup-and-release system on several F-86s so they could tow the aerial gunnery banners. We launched the first gunnery tow mission and four-ship gunnery training flight at Suwon Air Base after the war.

The Operations Cycle

We had three squadrons in the 51st FIW—the 39th FIS, 25th FIS, and 16th FIS. Each squadron had a three-day cycle, which covered the alert, training, and coastal reconnaissance missions of the 51st FIW:

The Alert Day

One squadron would always have two four-ship flights sitting near the end of the runway, ready to start engines and take off within three minutes. One flight would have pilots in the cockpit and power on the aircraft so they could take off within three minutes after the alert horn sounded. Two blasts on the horn scrambled two aircraft and four blasts scrambled four aircraft. The alert was a 12-hour duty, with each flight rotating cockpit duty every hour. We had the enemy intercept duty during the day. The F-94 Squadron on the other side of Suwon AB had a pilot and a radar operator and pulled the night and all-weather alert duties. When a flight was scrambled, the backup pilots would run out of the alert shack, climb into the cockpit and hook up, and the ground crew would connect their ground power unit to the F-86. They would be the next flight to be scrambled. Most of the intercepts would be unidentified targets in or near the demilitarized zone, except for two scrambles that I will never forget. More about that later.

The Training Day

The next day would be a training day and we would fly gunnery training flights, firing live ammunition from one of the six 50-caliber machine guns on air combat missions. The practice gunnery flights were just like the ones we flew at Nellis AFB in Las Vegas. One F-86 pilot would take off, towing the target banner 2,000 feet behind the aircraft and fly to the live gunnery range. He would fly at about 250 knots airspeed at about 15,000 feet. The four-ship flight would take off and join the target ship at about the time he arrived at the live gunnery range. Each F-86 would make an individual pass at the towed target, starting up on a perch about 1,000 feet above, and slightly behind and to the right of, the target. They would initiate a sharp-diving steep left turn, followed by a reversal to a steep-diving right turn to close on the target. They would start tracking the target tow line with the gunsight pipper at about 4,000 feet, flying the aircraft so that they could slide the pipper down the tow line. Hopefully, the pipper would be on the target just as the range to the target was 1,000 feet, the range at which the gun was boresighted. If you squeezed the trigger at 1,000 feet for a little less than one second, your bullets would hit the target. Your aircraft would be in a steep-diving right bank, and you would pass behind the target at a closing speed of about 100 to 150 knots. It was an attack more appropriate to kill bombers, but it was a very good drill for teaching the geometry and coordination required to get your target into the small kill cone of your fixed guns. The high rate of fire from our 50-caliber machine guns greatly increased the probability that several rounds would hit the towed target and, hopefully, enemy aircraft.

The Mig had two very slow firing cannons that fired larger and more lethal rounds and was designed to intercept and kill large bomber aircraft (B-29, B-50). In high speed, high "G" maneuvering combat, the six 50-caliber machine guns on the F-86 provided a nearly continuous line of lethal metal and gave us a valuable advantage over the enemy's slower rate-of-fire cannons. (Conceptually, we could fly through their line of fire and not be hit.) We needed advantages against these lighter interceptor aircraft, which could fly higher and turn tighter. The F-86 advantages were: We could out-dive them (supersonic straight down), the lethality of our guns, and a better flight control system with a superior rate of roll and controllability at lower altitudes. These advantages, and much better training, gave us our ten-to-one kill ratio over the Mig-15.

Dangling cigarette, two guns, and my Al Capone image

The other training activity that all fighter pilots love to practice is air-to-air combat against any other fighter flights flying in our "free for all" air combat exercise area. It was a large area in South Korea dedicated to flight-on-flight tactics, where the victor and his wingman end up with gunnery film of the victim, sufficient to convince himself that he would have shot down the victim, i.e., inside 1,000-feet range with his pipper on the victim's aircraft. (Our lead computing gunsight projected a circle with a dot— pipper— in the center on the forward windscreen, which was driven by a computer that revealed where the bullets would be at 1,000 feet, regardless of the maneuver attitude or "G" load.) Our gun camera line of sight was just below the pilot's line of sight, looking out through the windscreen with a field of view sufficient to cover the areas of interest for fixed guns. Whenever you pressed the trigger, the gun camera was turned on. If your guns were armed, they would also show the tracer rounds fired by your machine guns and the actual hits on enemy aircraft. When your guns were not armed, you only saw the gunsight on your windscreen and whatever was in front of your aircraft. You could infer range to target, because your pipper was two mils in diameter, which is two feet at 1,000 feet. Thus, we had a way to check our friendly kills by assessing our dogfight gun camera film.

All of the friendly fighter wings in Korea knew where our air combat area was and would go there looking for an air battle. Flights of four aircraft from the four USAF wings in our area, the USMC Fighter Wing, the South African

Attack Wing, and occasionally USN Carrier-based wings would come into our area in combat formation looking for a fight. We would be above 45,000 feet, scanning the airspace with our eyeballs to gain the advantage of first detection. The sighter would declare, "Bogies at two o'clock level," and the flight lead would commence his planned attack. All of us would put our power to 100 percent, and the air battle would commence. In actual combat the flight would also jettison their drop tanks as soon as they knew the sighted aircraft was enemy. The worst case situation is when you have been spotted first and one of your wingmen calls bogies coming up behind your flight.

The flight lead and element lead are the offensive shooters and scanners, principally in the forward hemisphere, and the wingmen always staying with the leader and scanning the rear hemisphere. Once you are engaged in battle, each flight generally splits into two, two-ship combat elements consisting of a shooter and his wingman, and the two elements try to attack in a scissors,

coming up behind the enemy from his right and left sides. If the enemy first spots the element attacking from his right rear, and commands his flight to break right into the first attacking flight, then your element attacking from his left side follows him into the right break and will be inside his turn and have an excellent shot at the enemy one at a time. After the initial attack, the air battle is played out as individual two-ship elements, with the element leaders trying to kill the enemy and keep track of where the other element is located. Once a four-on-four dogfight starts, every flight in the area joins into what is now a gaggle of elements all at full power pulling max "G's" trying for snap shots, diving to escape, and finally calling bingo fuel (just enough to return to the base and land). It is wonderful training, very realistic, addictive, and loads you with adrenalin.

We had trained at Nellis AFB to return to base low on fuel, because all of our instructors who had completed 100 missions related to us how they would fly all the way up to (and probably across) the Chinese border, jettison their fuel tanks, and engage in hunting for, and fighting with, Migs—with one eye on the fuel gage. They knew exactly how much fuel was required to get home and tried to disengage at that point. They would gather their flight while they climbed to 45,000 feet. If all flight members had enough to make it home, they would return to their base. Sometimes one or more of them were below the minimum fuel required. Depending on the amount of fuel left, they could: 1. land in Seoul Korea (but that was fewer than eight minutes from home; 2. turn off their engines over North Korea when they reached 45,000 feet and glide south (that gave them up to 80 free miles), restarting their engines when the fuel remaining was sufficient to make it to Seoul; or, 3. land on the sandy beach on Cho Do Island about three miles off the coast of North Korea. We held the island and the beach was on the far side, sheltered from North Korean artillery barrages. They had JP-4 fuel in barrels, which had to be pumped into the F-86 wing tanks until there was enough to take off from the sandy beach and reach Suwon AB. At Nellis I learned not to be excessively anxious about sweating out low fuel. As far as I was concerned, these were great contingency plans, but I did not plan to use them. All of those options were used by F-86 pilots during the war, but none was used after the war ended. The F-86 fuel gage was only accurate within 300 pounds of fuel. The training day was the most enjoyable and very valuable to maintain our combat edge.

The Coastal Reconnaissance Day

Every day a squadron (four four-ship flights) from the 51st FIW would take off from Suwon AB, and another squadron from the 4th FIW would take off from Seoul AB. One squadron would patrol the east coast of North Korea, with the other patrolling the west coast. We would fly to the southern border of North Korea and patrol three miles out to sea, climbing up to our maximum altitude (48,000 to 50,000+ feet, depending on the day) along the coast of North Korea with one squadron flying up to China on the west coast and the other flying up to Russia on the east coast. Then we would turn around and return over the same route. The North Koreans claimed their boundaries went 12 miles out to sea; we (the UN) only recognized the international law that establishes three miles out to sea as the limit for national sovereignty. Since they considered us inside their borders, they would occasionally send up several flights of Mig-15s to intercept us when we were about halfway up their coastline—they would always be about 2,000 feet higher alongside us, and between us and their coastline. When we reversed course at the northern border to go home, we would make our 180-degree turn out toward the sea to stay over international waters. The Migs always had problems maneuvering their 180-degree turn while retaining their advantageous altitude and staying with us. If they turned inside of us, they would appear to be attacking us, in which case we might defend ourselves vigorously. It was always the most inter-esting part of our mission. We all had to carry our largest fuel tanks because it was a long-range mission and the tanks gave us a lot of aerodynamic drag. If we dropped our fuel tanks and engaged in combat, we were a long way from home and would have little time to fight. With all of these disadvantages, we still wished they would give us an opportunity to defend ourselves. We were spoiling for a fight.

During the year that I was there, only two attacks were initiated by the Migs; and, of course, one Mig was promptly shot down in each encounter, which ended the battle. Both of these attacks were against our competitive 4th FIW. The 4th FIW shot down two Migs in postwar air skirmishes. We—the 51st FIW— had none because we never were attacked. Both of these skirmishes occurred about halfway through my tour.

It's a bird! It's a plane! ... No, it's Superman!

We analyzed the encounters and sure enough, one attack was initiated against the 4th FIW during the 180-degree turn at the northern end of the patrol. The other attack was against a USAF RB-57 flying a reconnaissance mission below the F-86 patrol that was flying high-cover for the RB-57. The F-86s promptly dove down and shot down one of the Migs attacking the RB-57. We never knew that the 4th FIW were also flying high-cover for reconnaissance flights. We did know that the North Koreans deeply resented these flights. Our squadron commander, Lt/Col William Beale, made numerous requests to obtain reconnaissance escort flights for our squadron but to no avail. The next time our squadron was assigned a west coast patrol, our commander had a plan. Both Mig attacks were on the west coast. Our ground radar controllers always kept us at the three-mile limit and reported any enemy aircraft activity. The Migs were given vectors to intercept our flight and any instructions to attack by their ground radar controllers, so both sides were watching the scenario very closely. Our squadron commander had Wes Shipstead (my aviation cadet classmate) and I fly close formation to appear as a larger aircraft at a lower altitude. We maintained radio silence so that the enemy radar controllers would think we were an RB-57 being escorted by the high-flying F-86 formation. It did not work, nor did any other ploys that we could think up to entice them to attack us. The coastal patrols offered us the opportunity to see many enemy Migs flying beside us in an advantageous position. We were carrying the battle to them and knew they would not venture into what we considered UN territory. That is deterrence through air superiority.

Thus, I spent my year in Korea gaining a lot of experience and progressing from wingman to element lead, and finally to flight leader without ever having the opportunity to apply these skills against the enemy. It was an honor, though, to serve under our leaders, to associate with the best of my peer group, and to be part of a truly professional military team.

On the Home Front

Sabra lived with her parents and family in Pensacola and ran her father's Bayou Vista Hotel and Restaurant. We both wrote each other every day. We wanted many babies and hoped and prayed that she would be pregnant, but we realized after I had been gone for several weeks that we were not pregnant yet. I sent most of my paycheck home and kept only enough for poker, cigarettes and beer. Her dad paid her a small salary and she accumulated a nice nest egg during my year in Suwon, South Korea.

Before the war ended I broke my right ankle playing touch football and went to the hospital to have a cast put on my leg. I wrote Sabra, describing how I had injured my ankle, and the next day wrote her again and told her I was in the hospital to have the cast put on. She received the second letter first and had no idea why I was in the

Pete, with Sabra as manager of the Bayou Vista Hotel & Restaurant

hospital, what was broken, and if there were any other injuries. She and her whole family were very upset and called the post office. They explained their concern and asked if there were any other letters from me. There were, and their friendly postman drove immediately to Sabra's home, delivered the letter, stayed until she read it, and rejoiced with the family over the relatively wonderful news. Those days are gone forever.

Our letters were basically the news of that day, followed by loving phrases and how much we missed each other. We saved them, and many years later found them, read and enjoyed them once again, then burned them. Now I regret that burning very much.

Memorable Missions and Events

Close Call

We were sitting cockpit alert with about four more hours to go before sunset. We had 12-hour shifts from sunrise to sunset, with two flights rotating one-hour shifts totally connected in the cockpit, while the other flight rested in the alert shack. The weather was bad, with light rain coming out of a dark stratus-overcast sky. We normally read books in the cockpit, waiting for the scramble horn to send us off on an intercept. Since it was raining, I had my canopy cracked open about an inch to keep the rain out, but open enough to circulate air to keep my breath from fogging up my canopy. Our hour in the cockpit was almost up, and soon the other flight crew would come out of the nice warm dry alert shack, strap themselves into the cockpit, and it would be our turn to read in a warm dry comfortable chair in the alert shack.

Suddenly, the saxon horn emitted four blasts—scramble four—our whole flight sprung into action: Put the book on the floor, push the start button, advance the throttle to idle, fasten your oxygen mask to your face, advance your throttle enough to follow your leader (who makes a high-speed, 45-degree turn onto the runway), then full throttle, and you take off on the wing of your leader. The leader contacted the ground controller and received a vector to our unknown target. As he turned toward the vector, the two-ship element (which took off just a few seconds after us) turned inside of us and joined into a tight four-ship formation. We stayed below the clouds until the element had come into formation, and then the leader climbed into the clouds, always receiving target information from the ground controller and following the vectors as directed by our controller. We came out of the clouds at about 33,000 feet, still climbing. I began to notice a vision problem and rapidly noticed that I was experiencing tunnel vision—my field of view was becoming narrower and was closing in on me. Suddenly I knew I was just a few seconds from being unconscious, perhaps forever.

We had received intensive training on the danger of hypoxia—becoming unconscious from lack of oxygen. Every year all high-altitude pilots would go into an altitude chamber; air would be pumped out of the chamber until it was equivalent to 35,000 feet. We would have our oxygen masks on and would be receiving oxygen under pressure sufficient to keep our brains fully functional. Each of us would be assigned a partner. One partner would take off their oxygen mask and try to count to 10 while the other partner watched closely and listened. They would count to three or five at the most, before becoming incoherent. Their partner would place their mask back on their face, and after five or ten seconds they would open their eyes and shortly afterward begin to talk. Then the partners would reverse roles. I vividly recalled my impressions as I was going unconscious. I remember counting all the way to ten (My partner said four.). I remember feeling very happy or giddy and my vision narrowing down all the way to darkness. Had my partner not intervened, I would have gone completely unconscious, unable to fly my airplane, and probably crashed. I learned that my perceptions of my performance were much greater than my partner's perceptions of my actual abilities. That is why they repeated this training every year for those of us who flew at high altitudes.

The training paid off. I pulled my throttle back, extended my speed brakes and turned my oxygen to full-pressure breathing. I thought I also made a radio call to alert my flight leader of my emergency, but they all denied receiving the call, which meant that my brain intended to call but failed to execute. My actions, however, resulted in a rapid but uncontrolled descent down through the clouds. I remember waking up, and I was in an airplane below the clouds and nearby mountains in an unusual attitude. Quickly, I took control of my F-86, leveled off, and reported back to my flight leader. Wendell Wood had come with me, returned my call and joined on my wing. The leader had completed our intercept on a friendly aircraft and we were all to return to base. They were still at high altitude, and Wendell and I were at low altitude beneath the cloud layer.

On the way home I analyzed my situation and found two related fatal mistakes. A single-engine fighter pilot usually makes only one fatal mistake per lifetime. I had failed to close my canopy fully (Remember, I had it open an inch to keep the rain out of the cockpit.) and my oxygen mask was not fitted tightly enough to keep the mask from leaking under pressure-breathing. The open canopy prevented my cockpit from pressurizing. Normally, when I was at 34,000 feet with my canopy closed, my cabin pressure would have been about 25,000 feet where pressure-breathing is not required. With my canopy open an inch, my cabin pressure was actually 34,000 feet, where pressure-breathing is required. My oxygen system compensated for this error because it sensed the actual cabin pressure as 34,000 and supplied my mask with sufficient pressure to provide oxygen to sustain high performance. However, my mask was not tight enough to keep the pressure at the required level. My mask was a comfortable fit but totally inadequate for the high altitudes where we operated, particularly if cabin pressure was lost. I had about 10 seconds to recognize my symptoms of hypoxia and take corrective action. These two errors should have killed me and were totally my fault. The excellent training, which taught all of us to recognize our hypoxia symptoms in time to take corrective action, saved my life. I vividly recalled that when I regained consciousness I was below the surrounding mountaintops in northeast South Korea. I learned that if I were to pursue this profession, attention to every detail was essential, and all details had a purpose. I returned to base, related my lessons learned, and ate a lot of humble pie.

Only Enemy Encounter

Our flight was sitting in the cockpit on alert on a bright sunny day several months after the war had ended, when the scramble horn sounded four times. The four of us were off the ground well within the three minutes allowed and vectored north by a very anxious controller who had a fast-moving radar target coming out of North Korea toward Seoul. Seoul, the capital of South Korea, was only 30 miles from the North Korean border. The radar controller had also scrambled four F-86s out of the 4th FIW in Seoul. As they were taking off to the north, the flight leader saw a jet landing toward them. A head-on collision appeared to be imminent. The first two F-86s barely got off the runway, flying over the oncoming jet that was already on the ground. The second element of two F-86s saw the near head-on collision ahead of them but did not yet have flying speed. They separated, and the landing jet went between them. All four F-86s got safely off the ground, and the lead reported (to the tower) the idiot who had nearly collided with the entire flight. They then contacted radar control for their vector to intercept the unidentified target. By this time, our flight was already over their airfield scanning the skies for our unidentified target, which had disappeared from the radar scope.

Meanwhile, on the runway below, the jet that had landed to the south was spotted by the alert crew. They assumed that one of their F-86s had a runway abort and was returning to the alert pad to park. They gave him hand signals and guided him into one of the recently vacated alert spaces. The relief alert flight were strapping themselves into their cockpits when the flight leader saw a Mig-15 park next to him!!! His radio call to the tower brought a flurry of activity. The wing commander quickly had the North Korean pilot, Kim No Sue, brought into the headquarters and the Mig-15 towed into a hanger out of sight from enemy eyes. Our higher headquarters ordered our two airborne F-86 flights, which had scrambled to intercept this intruder, to maintain an airborne alert in orbit at high altitude over Seoul to intercept any other enemy aircraft that might join this mysterious invasion. We orbited until we were low on fuel and relieved by two other flights of F-86s to continue the airborne alert.

Kim No Sue was interrogated as soon as a Korean translator arrived. He claimed that he was a North Korean defector who wanted asylum. The USAF had dropped leaflets over North Korean airfields, offering $100,000 for an intact operational Mig-15, and they now assumed that this was his motivation. Kim stated he had never heard of the reward and his only motivation was seeking freedom. He was an attack pilot flying many combat missions from his airfield in North Korea until they were moved into the safe haven of Manchuria inside China. Then he was trained by Russian pilots to fly the Mig-15 aircraft against the F-86. The Russians always had a fighter interceptor group at this airfield and both groups flew high along the Manchurian/North Korean border looking for UN aircraft to kill in North Korea. They probably were forbidden to fly over North Korea because a Russian POW would be a political disaster for the USSR. Kim estimated that at least half of the missions flown in Manchuria were by Russian pilots, and they had claimed most of the kills. They probably accounted for most of the F-86 versus Mig-15 air battles. Both the Russian and USAF pilots were eager for battle and anxious to find the strengths and weaknesses of their aircraft, pilots, and tactics.

Kim and several other North Korean pilots were trained by the Russians to fly a ground attack mission from their Manchurian airbase against the

Kimpo Airbase in Seoul, where our 4th FIW was stationed. The Russians carefully planned their flight to fly high to conserve fuel, then descend to avoid radar detection, flying very low and attacking Kimpo at low lever. It probably was a oneway mission because of the limited range of the Mig-15. The Mig pilots who survived the attack on Kimpo may have had enough fuel to return to their side of the border and eject. The plan was scrubbed after hostilities ceased, but Kim had kept a copy of the flight plan. His squadron's political officer became suspicious of Kim and grounded him for a short time for re-indoctrination. This made Kim more anxious to escape, and he eagerly awaited an opportunity. When the opportunity arrived, he executed it perfectly until his downwind landing, head-on into a flight of four F-86s. The USAF finally had an operational Mig-15.

Kim's story was credible, and the US government gave him his $100,000 reward, concealed him in the new-identity program and wished him well. I met him more than 40 years later, when he started attending our 39th FIS reunions. He had used his reward money to learn English and earn an aeronautical engineering degree. He retired as a professor from Park Air College. He was reunited with his mother and wife somehow and has written several successful books relating his adventure. I am very happy that we failed to intercept him before he landed at Kimpo. We all enjoyed one another's company. It really makes you realize how lucky we are to have our freedom. Kim had survived the tyrannical rulers in North Korea, fought us as a fighter pilot, and survived a truly risky escape—all to enjoy his freedom. I am certain he truly appreciates freedom, a gift we take as granted.

The Mig-15 was loaded into a C-124 and flown to either an airbase in the Philippines or Okinawa. A number of test pilots calibrated its performance, and several Mig aces flew the Mig-15 against the F-86 to determine the relative strengths and weaknesses. This provided very valuable data to develop tactics for use in fixture air battles. The Mig-15 was flown extensively until it was retired to be displayed in the USAF Museum at Wright Patterson AFB in Dayton, Ohio, alongside a photograph of Kim No Sue and a brief version of his story.

Each flight was assigned a house boy to keep the flight barracks clean and wash flight suits and bedding for the 14 occupants of the D-flight barracks. Our house boy was Kim (a very common name for Korean males). He was short and looked very young, but he worked very fast and efficiently. I think the other house boys wished he would work slower because they all looked slow compared to Kim. We all were very fond of Kim, and he learned a decent amount of our language, mostly from Bob Groszer, who took him under his wing. I can still visualize Kim's smiling face and his baggy pants and baseball cap.

The Moon Suit Experiment

The water off both coasts of North Korea was very cold in winter. If our aircraft were damaged and we could not make it back to friendly territory, we would try to make it out to sea. An ejection over North Korea would probably result in an extended stay in captivity—even a very cold sea looked better than captivity! If we ejected and parachuted into the water, we would lose most of our body heat before a helicopter would be able to rescue us. The time of useful consciousness in that cold water was not very long. So they invented and bought all aircrew members at risk of parachuting into cold water ... a "Moon Suit."

The moon suit covered the entire body (including the flying boots) up to the neck and hands. It was made of canvas with rubber seals around the neck and wrists, and it took a long time to enter and exit this contraption. We wore it all winter on all flights until the water temperatures were survivable. On alert days we would spend more than 13 hours encased in this unventilated pimple factory. Since a fighter pilot always has his head on a swivel as he scans both the forward and rear hemisphere, we each developed a red, chafed neck where the rubber seal pulled against us as we swiveled. We claimed that the suit may save us from the cold ocean but made it more likely that we would not see enemy Migs attacking us from the rear, because the suit reduced our rear-scanning ability. So, our local machine shop made a metal ring to insert in the rubber neck seal to keep the rubber off of our chafed necks, and that solved that problem.

The other problem was that with 13-hour workdays, provisions had to be made to urinate. The designers accommodated that function by providing a tightly zippered entrance, which contained a rubber hose to facilitate urination. They must have asked drunken fighter pilots in a bar to determine the dimensions of this hose. It was at least twice the dimensions of the legendary Hercules. They should have queried the wives. I doubt if anyone ever managed to urinate via that tube in flight (The F-86 was equipped with a pilot's relief tube.). It was very difficult to perform this task, even out of the cockpit. Each pilot apparently solved this problem on his own. My solution was to cut the tube down to a size where I could relieve myself while on those long 12-hour alert days, when I was out of the cockpit. There was no way I could perform this task while fastened into the cockpit, and I never tried.

The claims that the discomfort of the moon suit decreased our combat performance persisted, and the Wing decided to run a test to determine the actual value of the moon suit in extending our survivability. The test was to drop about ten of us into the Yellow Sea in our moon suits to see how long we could function in the cold water (hopefully long enough to be rescued by a helicopter). They selected pilots who had been in the theatre for varying amounts of time to determine if wear and tear impacted the durability of the suit. Cho Do Island off the coast of North Korea was the test site, which was the northernmost location with the coldest water.

We flew to Cho Do in a C-47 transport plane, all of us wearing our moon suits, and landed on the beach. Two helicopters awaited our arrival. We received our briefing and were flown a short distance out to sea, where a small rescue boat was orbiting. We were lowered a few feet into the water, with only our hands and faces exposed to the cold water and wind. We always wore several layers of clothing under our suits, plus an inflatable life vest, so this is the way we would arrive in the water when we ejected. Only our inflatable life raft was missing, since we wanted to test the durability of the moon suits.

I was surprised how buoyant I was with all of the trapped air in the moon suit ballooned up around my upper body. I was so buoyant, in fact, that I had difficulty keeping my feet down and did not even inflate my life vest. The cold impacted my face and hands immediately, but my body was reasonably warm. The seas were light, and I could see most of my fellow floaters. Soon I noticed that several were going lower in the water and then inflating their life vests as they lost buoyancy. Suddenly I understood their problem, as I felt the cold water coming in, in the vicinity of my crotch. My idea to trim down the size of the relief tube in the suit had compromised the seal, and I was slowly but surely sinking. However, I was the new guy with the newest moon suit, and my comrades were sinking faster than I was.

The helicopters and the boat were very busy, and I was the last one picked up. I was in the water the longest, but I have no idea how long it was—probably about 20 to 30 minutes, and probably not long enough for a rescue. There was no panic. Perhaps the cold water dulled our senses. The temperature inside my suit changed gradually because my body heat was trapped and the water leak was slow. My exposed face and hands were very cold. We were taken back to the beach and given warm clothes, and we flew back to our base. The moon suit was probably a good idea before the pilots made their own modifications to facilitate urination, but the wear and tear on the zipper and trimming down of the very long tube eventually led to a growing leak problem, as time and usage took its toll. A moon suit filled with water and a pilot weighs a lot. I wondered if the helicopter wenches could haul them out of the water.

I understand the moon suit was modified and use continued, because I heard the discomfort complaints from other pilots over the years. I never had to wear one again after my Korean tour ended.

Miscellaneous Events

Our flight commander, Major Jack Bolt, USMC, rotated back to the United States about two months after I arrived, and we all eagerly awaited the arrival of his USMC replacement. He arrived after dark and came into our hut carrying all of his gear. We all gathered around and began introducing ourselves. He started, "I am Major Erskine B. Crew, USMC." ... A voice from my past. Captain Erskine B. Crew had been my tactical officer in charge of our USN Midshipman Pilot Training Pre-Flight Class just four years before. This was the Marine who had wielded the power to eliminate me from the Navy for sixteen weeks by giving me one more demerit. I had been awarded the maximum allowed demerits for unforgivable behavior before pre-flight, and he had vowed to "watch me like a hawk for the next sixteen weeks." He did; and I made the sixteen weeks without a single demerit—not missing a salute, shoeshine or haircut, without a wrinkle in my uniform or bed in a room devoid of dust, and rigorously controlling my mouth. He used to put on his white gloves when he entered my room for inspection. When I graduated from pre-flight, he congratulated me and told me he was proud of me for my effort. I never knew anyone else who made it through pre-flight without a single demerit. I also never knew anyone who was given 50 demerits by a board of officers before they started their Naval career.

I approached Major Crew and said, "Second Lieutenant Robert W. Kennedy, sir." He looked at me, shook my hand, and told me he remembered me from our days in Pensacola. He was my flight commander for the rest of my year in Korea. We slept in the same hut, ate, drank and flew together, and he treated me like all the other troops in his flight. He could have made it very difficult for me, because he knew it was illegal to go through flying training twice. I knew he would test my flying abilities thoroughly before I was off the hook and he did, occasionally flying my wing in weather and in our air-to-air hassles. I usually had the highest gunnery scores, and soon I knew I had his approval. At last I was vindicated from what I had always considered an unfair "washout." Both of us kept our memories to ourselves, and this is the first time I've revealed our secret.

Our 51st FIW had adopted an orphanage run by Korean nuns, which was located about five miles from our base. We supported it financially and personally. When we had an afternoon off, a jeep-load of us would drive out to the orphanage and play with the children. I would softly wrestle with as many kids as cared to leap into the fray. We could not communicate, but I loved to hear them squeal and make joyous noises. I would be on my knees, and they would attack me vigorously from all sides. I would grab them and gently deposit them upside down elsewhere, or push them away with my "panda paws." They were tiny and needed adult companionship and play. When they became tired, they would lie close to me, not quite a cuddle. When we rode back to the base from our orphan frolic, we would all be very quiet and sad. It felt very good to bring a little joy into their lives, but it was not an adequate substitute for loving parents ... and that made us sad. I looked forward very much to having many children with my lovely bride when I returned home.

We received one Rest and Recreation (R&R) during our tour, and Wes Shipstead and I went to Tokyo for about a week. I bought two strands of

Mikimoto pearls and a pair of earrings plus a tailored silk Japanese dress for my bride, and a camera and a wooden model of my F-86. I had my name painted under the canopy rail and "Sabra's Sabre" painted on the nose, an identical replica of my F-86.

We spent one night in the "earthquake-proof International Hotel for $10 (now over $500), ate Kobi steak, had a hot bath, and visited several nightclubs. Both Wes and I were married and did not fool around. Most of our lieutenants were bachelors and had a ball. We took a taxi up to the Fuji Golf Resort near the top of Mount Fuji. The Japanese taxi driver claimed to be a kamikaze pilot, but the war ended before he had his opportunity. We drove up those winding roads full down on the gas or brake, passing on curves—undoubtedly the most dangerous automobile ride of my life! Once again we were living in luxury in the hotel—sheets, towels, bed, pillows, all were very plush. We rented golf clubs and caddies who appeared to be very young girls and played 18 holes on the side of Mount Fuji. When I sliced or hooked the drive, my little caddy would run after the ball into the grass that was as tall as she was and always come up with the ball! We had a wonderful time for a very small amount of money. It was 1953 and an American dollar bought a whole lot of Japanese yen. Unfortunately, we rode back to Tokyo with another Japanese taxi driver, who claimed he used to be a kamikaze pilot, and going down Mount Fuji was more terrifying than going uphill.

Fritz Meyers and Jerry Carlisle chose a tour in Japan after they completed their Korean tour rather than returning to the States, and both of them were assigned to Johnson AB in Japan. Tragically, First Lieutenant Meyers, a member of our flight, flew his F-51 into Mount Fuji in the weather and died just a few months after our visit to Tokyo.

I learned a lot during my year in Korea because I was in the fighter business full time and always with my leaders and squadron mates as a close-knit team. I occasionally saw the enemy Migs flying alongside us on our coastal patrols but never had the opportunity to engage them. We had a lot of fun drinking and singing in the officers club. I made a lot of friends, whom I still see occasionally, even 50 years later. I departed Korea one year after I arrived, eagerly awaiting my reunion with my bride. Luckily, I had been assigned to the Air Proving Command at Eglin AFB Florida, just 40 miles from my bride's home in Pensacola. We had written each other every day—short, passionate love letters—and I vowed to tell her every day that I loved her. I kept that promise over our lifetime together.

I said farewell to my squadron friends and my armament troops, and they drove me up to Seoul to catch a C-124 to Japan. Upon arrival in Japan, I learned there was a 15-day waiting list to catch a military flight back to San Francisco. They offered a six-day trip on a military troop ship going to Seattle, and I volunteered immediately—I would be able to see Sabra 10 days sooner. They took us down to the port and we sailed that evening ... back to Pusan, Korea! There was a large group of Turkish troops in the hold going to Korea as replacements. I was the junior officer aboard, still a second lieutenant, and was assigned as acting commander of the Turkish troops. I had one Turkish sergeant who spoke English. He would meet me early every morning to inspect the troops. It was very obvious that Turkish soldiers had a respect for officers that bordered on fear. Those troops were known as the toughest warriors in Korea, and I was glad they were on our side. They disembarked in Pusan, Korea, and we sailed for Seattle. I will never forget the "Welcome Home" sign at the entrance to Puget Sound in Seattle. I had sailed past that sign when I returned from my three sealing expeditions; but now, at long last, I was coming home to my bride!

CHAPTER 10

Stateside

I took a cab from the Port of Seattle to the commercial airport, bought a ticket to Pensacola, Florida, and arrived to the arms of my bride about 16 days after we had planned our reunion. I felt whole again, reunited with the other half of my soul. Her parents and all of her little sisters and Phillip Joe were all there to welcome me home. We all went out to dinner at Martines Restaurant that evening to celebrate my homecoming; and her cousin, Father Henry Thorsen, said the blessing. Father Thorsen received his 'wings' as a priest the same week that I was awarded my Air Force pilot wings. It was a very heart-warming reunion.

I had called my family in St. Louis and my sister in Los Angeles from Seattle right after I called Sabra. We went up to St. Louis to visit my family and friends for about a week and then back to Eglin AFB to report for duty. I was assigned to the Base Flight Squadron as a line pilot and one of the many assistant operation officers. Our Squadron had 11 different types of aircraft assigned and I looked forward to flying all of them. But first we had to find a home.

Our Life Plan

I had drawn about $40 a month from my pay (about $350) and sent the remainder as an allotment for Sabra. She had run her father's Bayou Vista Hotel and Restaurant and lived at home. She made a profit for him that year and he paid her $500 per month with room and board. She had saved all of the money, and when I came home after one year in Korea we had enough to buy a home and set up our life plan.

After much searching, we bought a very small, two-bedroom, one-bath, square concrete block house (painted pink) at 657 Highway Boulevard in the Garniers Beach subdivision in Fort Walton Beach for $10,000. It was less than 45 miles from her parents' home in Pensacola. We had a 1952 Plymouth automobile, a bed, a couch, a home and adequate kitchen tools to operate our very happy marriage with great hopes for many children. We had made a list of objectives and a plan to achieve them in Las Vegas before I went to Korea, and of course our objectives were to have babies, own a home (never pay rent), purchase furniture and minimum household necessities, and pay off our car debt of about $1,300. We basically planned to pay cash for all acquisitions and limit our debt to our home mortgage where our home appreciation would more than offset our interest and principle expense. We made a budget, and the first of the month Sabra would deposit all funds that exceeded our budget into our savings account. We maintained this discipline throughout our military career and used our savings to buy furniture, a car when ours wore out, and any unforeseen repairs or emergencies. We kept a minimum savings reserve of about four months salary in case I lost my job, and made our furniture/auto/clothing purchases out of the excess savings. I remember it took us about three years before we completely furnished our very small home. After we had accumulated our furniture and paid off our car we began to invest our savings. We never carried debt other than our mortgage and we always sought maximum leverage on our mortgage; i.e., minimum down payment. Our home's total value would always appreciate 4-7% per year, a sum far larger than our down payment, to yield an excellent return on our down payment investment. We bought several homes during our career with zero down payment VA loans, and interest rates were 4-5%. A far cry from our modern newlyweds who buy a home, two cars and furniture all on credit shortly after they marry.

We updated our objectives and Life Plan throughout our whole married life, checking the objectives off as we achieved them. We executed the well-known and usually ignored advice to budget less than you earn and save and invest the excess every month. The 'magic' of dollar-cost-averaging investment made us millionaires during our sixties. There was no magic but a lot of planning and discipline.

Reporting for Duty

I was welcomed into the Squadron by our commander, Lt./Col. Glen Stitt, and Major Olin B. Carter, our operations officer. They assigned me a lot of additional duties since I was the most junior officer in the Squadron. It turned out I was the most junior pilot on the base! I reported in as a Second Lieutenant but was promoted to First Lieutenant after my promotion orders showed up on base. My 18 months in grade had been completed right after I left Korea, so I spent an extra month as a 2/Lt. However, they back-dated my pay to make up for it. I may be the only Air Force officer who spent 19 months as a 2/Lt.

Eglin AFB Florida was the home of the USAF Air Proving Ground Command, with the mission to perform operational testing on all new USAF aircraft, weapons and equipment. We received the second batch of new aircraft from the production line and put them through operational testing to include weapons testing. The tropical tests were done in Panama and the cold weather tests at Eielson AFB Alaska. The Flight Test Center at Edwards AFB California received the first batch of aircraft to test them against their contractual performance specifications and to define the flight envelope. While I was at Eglin, they tested the F-100, F-101, F-102, F106, B-58, and B-66. I was in Base Operations and we did all of the support flying for the entire command. It was excellent experience learning all about operations at a base with all of the new and old USAF aircraft.

Most of the pilots at Eglin had seen combat during WWII and I learned a lot from them. I had the Jet Training Upgrade Program with six two-seat T-33A Jet Trainers and one single-seat F-80C, identical to the T-33 except for the single seat. I had the privilege of checking out many of these very experienced combat pilots in jet aircraft. Most of them had only flown conventional-engine propeller aircraft. They all were delighted to transition into jet aircraft, and I learned more from them than they learned from me. I was fascinated by their WWII combat stories and (nearly) all of them were excellent pilots requiring only several hours of flying instruction into the unique aspects of jet engines and aircraft. They all loved the very responsive flight controls and the increase of power and speed. I was privileged to check out the Base Commander, and the Vice Commander and Commander of the Air Proving Ground Command, a Colonel, a one-star and a two-star General. I had never had more than a two-word conversation (Yes Sir) with any officer senior to my Squadron Commander who was a Lieutenant Colonel. They all were very friendly and professional, and soon I felt very comfortable as their jet instructor. I learned to be a professional instructor, never exploiting our relationship after landing and debriefing. I wrote a small publication, "Flying The T Bird," which helped my pilots new to jet aircraft with some of the differences between flying jets and piston-driven aircraft. It was well received because it revealed that flying jets was much easier than props and contained some hints that made the transition easier.

My boss, Major O. B. Carter, really drove me hard. He was not happy to have a very junior Lieutenant with only one operational tour in his unique operation and was determined to provide me with a lot more diversified flying experience in a very short time. He needed me because he had just received the seven jet training aircraft, but all of the rest of his aircraft were conventional propeller aircraft, circa WWII, and there were only seven full time pilots in his Squadron. We all were expected to fly these aircraft on myriad support missions, so he started checking me out and seeing that I logged enough hours in each of his aircraft to maintain proficiency. Within two years I was checked out in the two-engine C-45, C-47, B-25, B-26, SA-16 including water landings and takeoffs, C-131 and was co-pilot qualified in the four-engine C-54, and C-124. I also had 30 hours in the H-13 helicopter. It was wonderful flying experience and I loved every moment, knowing that this diversified flying was a rare opportunity.

We had more than 100 pilots attached to our squadron who had full time staff or command jobs, and they all had to satisfy their annual flight requirements—100 hours flight time including night and weather/hood, a flying proficiency check, and an instrument flight check and written exam. Our full-time Squadron pilots checked them out in their aircraft, gave them their annual flight and instrument checks, taught the instrument-flying classes, supervised the enlisted operations and scheduling personnel, and flew numerous test and logistics support missions for the command. That is the normal range of duties for a Base Flight Operation. That is normally a dead end job for a pilot. You learn a lot about running an air operation, but it is not directly involved in combat operations as long as we control the air above our bases. Thank God at Eglin AFB we had many other exciting and unique duties.

Training Operations

We also put on the USAF Airpower Demonstration twice a year for the executives of American industries, Congressmen, other VIPs and the media. They all would come for a two-day demonstration of the USAF capability. We would house them in our best available quarters, feed them, and put on an aerial exhibition of airpower with live munitions and high-speed passes by all of our future and current tactical and strategic aircraft. My good friend from Korea, First Lieutenant Red Iglehart, was assigned to the demonstration team to demonstrate a toss bombing delivery in the F-100 and was killed on a practice run the week before the demonstration. He pulled up into some cumulous clouds, and the next time we saw him he came straight down out of the clouds into the ground.

The formal demonstration went very well. It was a very impressive performance and I am certain that we enjoyed the full support of all of our invited guests. I was present for all of the support planning involved and also directed the jet indoctrination flights for carefully selected VIPs in my six two-seat T-33 Jet Training aircraft. Our six Instructor Pilots would give the selected VIPs a 3 5-minute flight after strapping them into the rear cockpit and giving them a briefing on the flight and emergency procedures. Throughout the flight we kept them informed about all of the visual, aural and maneuver events before they happened. I never had a passenger that became airsick during one of my orientation flights. If they told me they were feeling a little sick, I would give them control of the aircraft and let them fly so that they could relate the motion that they were seeing and sensing to their own control stick/throttle movements. Almost all of them would be able to perform an aileron roll and a sharp turn pulling some Gs before landing. They were very impressed and enjoyed it immensely. A photographer would meet us at the aircraft and take a picture of the two of us in our flying gear after the flight. Most of them wrote me a nice letter when they returned home, telling me how much they enjoyed the flight. One of my passengers was the President of Monsanto Chemical Company, Mr. Sweeney, and he was elated to meet the grand nephew of his long-time executive administrative assistant Eva Cross. (See pages 4, 5).

We also provided summer training for the Annapolis and West Point cadets as well as some Air Force ROTC cadets. There was no Air Force Academy in those days, and the Air Force was allowed to accept up to 10% of the Annapolis and West Point graduates. I ran the water survival and T-33 orientation program for all of these summer visitors and did my best to inspire them to come into the USAF. We could tell which ones were most impressed by their reaction to the jet orientation rides, and I believe we played an important role in their decision process of selecting their service or a flying career.

The water survival course was minimal instruction and a lot of fun for all concerned. We had several PT boats and two SA-16 amphibian aircraft assigned to Eglin AFB because we had a very large portion of the Gulf of Mexico assigned to us as a weapons test range. These resources would normally be assigned to the Coast Guard, but we owned them and used them to clear the ranges before any weapons testing. Often stray fishermen who had failed to refer to their charts, which clearly showed our ranges as danger areas, would wander in and we would have to lead them to safety. We also used these PT boats for water survival purposes. We would put cadets overboard in groups of six with a six-man life raft and they would have to inflate it, climb aboard, access the emergency water, and exercise the signaling devices (flares, whistles and reflecting mirrors) and shark repellent. We would then repeat the exercise in one-man life rafts. The Gulf waters were always warm and everyone had a good time, except for those who could not swim or had a fear of the water. They were referred to their own units for further training or elimination.

I always loved to fly the SA-16. I had a very hard time taking off in water. You had to apply unequal power when you started to lift one of your outboard floats out of the water, being careful not to over correct and put the opposite outboard float into the water. Then when both floats were out of the water and your engine throttles were at full power, you had to get your aircraft up on the step into takeoff attitude until you had enough airspeed to rise into the air. It sounds simple but it required a great deal of coordination and concentration. After several attempts I finally got it altogether. Water landings were very easy except for open water landings in the gulf. We used to practice those occasionally until we found out that the SA-16 had a tendency to nose into a wave and sink if you landed in the trough of a wave in a medium sea state. When we finally had to transfer our SA-16s to the US Coast Guard, we learned that our two aircraft had the most open water landings of any of their SA-16s. They considered it an emergency procedure to be avoided if possible.

Cold Weather Experience

The Command also had a cold weather hanger on base and a Detachment at Eielson AFB in Alaska to perform cold weather operational testing on new aircraft being introduced into the USAF. The cold weather hanger at Eglin could test our aircraft and weapon-maintenance procedures and engine start-up and operating procedures down to 30 degrees below zero. The new aircraft would be flown to Eielson AFB in Alaska in November and fly operational missions throughout the winter at temperatures down to 55 degrees below zero.

Our Squadron flew all of the support flights between Eglin AFB Florida and Eielson AFB Alaska, near Fairbanks, using C-47 aircraft, the cargo version of the 1936 Airline DC-3, fondly known as the "gooney bird." I worked very hard to become qualified as first pilot so that I could fly one of the Alaska support missions. As a youth I had been on three summer expeditions as a seal blubberer in the Pribilof Islands Alaska but had never been there in the winter. I badly wanted really cold weather experience. At the last moment in mid December 1954 I finally had my opportunity, and I was rewarded with a really bad cold weather experience.

We had an unscheduled requirement surface in early December to make a supply run to Eielson to deliver critically needed supplies before 17 December. None of the attached pilots would take the flight because any bad weather delay would mean that you might not make it home before Christmas. I jumped at the chance, and one of the bachelor attached pilots volunteered as copilot. I picked two of our most experienced enlisted crew chiefs who had made the trip several times and were familiar with the very cold weather maintenance anomalies. It was about a 2700-mile flight one way. We planned to refuel twice and spend one night in Rapid City AFB South Dakota and one night at Whitehorse Canada, flying into Eielson by noon of the third day so that we could unload and load the return cargo. We could return home in two and one half days, arriving December 21st, the day before our wedding anniversary and my birthday. That gave us some slack in our schedule, which we could make up by flying longer each day.

All went as planned on the route until about an hour from Eielson, when we entered a strong cold front and clouds with severe icing. The terminal weather forecast was "low clouds and visibility," but we had an acceptable alternate airport. In December the sun is visible less than one hour per day, so basically all of our time in Alaska was nighttime. Since the forecast temperature at Eielson was 55 degrees below zero, we had reduced our tire pressure to compensate for the temperature differential between our take-off and landing bases. We had turned on all of our deicer equipment but still loaded up on ice once we got into the clouds and had to add power. When we started our approach at Eielson, we lost the right engine and could not get it started so we feathered it to reduce the drag from our windmilling prop. We were very heavy with ice on the entire aircraft and had a moderately heavy cargo load. The weather was about 300-foot ceiling and ½-mile visibility, so we needed to make a Ground Controlled Approach with our controller giving us verbal directions to keep us on the glide slope and runway approach azimuth. When I put down our landing gear and flaps, I could not stay on the glide slope with full power on the left engine. The weight of the ice and the distortion of our wing airflow caused by the buildup of ice behind our deicer boots on the leading edge of our wings caused us to require more power to keep the heavier airplane on the glide slope. The disruption of the airflow increased our stall speed, so we had to increase our velocity to avoid stalling out and crashing. I retracted the landing gear to decrease our drag so that we could stay on the glide path, and the copilot desperately looked for the runway. As soon as we spotted the runway we dropped our landing gear. It went down and locked just as I started my round out for landing. As soon as I pulled the nose up to round out for landing, we stalled out and hit hard on the runway. I was carrying about 15 miles per hour above normal approach speed and the ice stalled me out way above our normal stalling speed. We went to the end of the runway and spotted the follow-me truck to guide us to our parking space. We had our artic flying gear on because it was 55 degrees below zero. With only one engine running we had very little heating power inside the aircraft and our breath condensed on the very cold windows. We went through our cold weather shut-down procedures and opened the door to a blast of the coldest air I had ever experienced. Although the landing was hard there was no damage to the landing gear. I was very happy to be on the ground safe and sound!

We had to have a new engine flown in and installed, so I not only missed our anniversary but also Christmas. My aircrew and I celebrated Christmas in the Eielson AFB chow hall. This was the second of our anniversaries that I had been separated from my beautiful bride, and I had volunteered for this one just to get cold weather experience! I called her on our anniversary and Christmas and told her how beautiful the northern lights were—like a continuous light show in the northern sky when the clouds did not obscure the show. She was only slightly impressed, and I decided that, hereafter, I would wait for my flying experience to accumulate without volunteering for risky flights near our special days. We finally had our engine installed and flew home to Florida arriving on New Years Eve. Sabra and I enjoyed a wonderful New Years Eve together. I had learned a lot about cold weather flying—the hard way!

Professional Education

The Air Force selected promising junior officers early in their career to receive additional professional education. The first course was Squadron Officers School at Maxwell AFB Alabama, and I was selected and attended this four-month course early in 1956. The class was composed of Junior Air Force Officers from every functional discipline and every Command. All of us were competing in leadership, athletics and academics. We generally had a half-day of lectures by very experienced combat leaders, politicians, academia, ambassadors, journalists and philosophers, followed by a discussion period within our own smaller group of about ten people. The afternoon was spent on athletic competition between groups, and field exercises that always involved resolving physical problems as a group. Our group leader observed us and graded us on athletic ability, leadership and academics during the four months. One week was devoted to being captured during a field exercise, placed in a very realistic Prisoner Of War camp, deprived us of sleep and adequate food, and finally allowed us to escape and try to evade recapture. They basically kept us stressed and under pressure to simulate a POW experience while observing who could tolerate hardship and emerge as teammates and leaders. We all had to write a term paper regarding the future USAF. My paper exploited the very recent development of nuclear power, which I envisioned could be used to extend the three dimensions of airpower, airspeed, altitude and maneuverability, to include a fourth dimension—endurance. It consisted of a fleet of large amphibious airborne aircraft carriers powered by nuclear reactors that could remain aloft for several days and could launch and recover manned and unmanned interceptors, reconnaissance and attack vehicles. They would be sea-based and capable of rapid response worldwide with a long loiter time in the battle area. The battle staff and logistics for several days of engagement would be on board, and restocking would occur at sea. My advisor was impressed and told me it was a great idea for the US Navy. It was a very enjoyable experience and I learned a great deal about our military doctrine, national affairs, aggressive competitive sports, leadership and myself. Sabra and I lived in a small cheap apartment in Montgomery, Alabama, with many other families attending the same course and we all had a wonderful social life during the four months. As usual, we formed many lasting friendships.

We spent another two years at Eglin and I was awarded a fully qualified specialty code as Air Operations Officer, supervised the Jet Upgrade Training section, instructed instrument school and water survival and was an Instructor Pilot in the T-33 and C-47 giving annual flight and instrument checks to the assigned pilots. During my five-year tour I flew 11 different aircraft, accumulating 2000 flying hours. I was the most junior flying officer in the entire Air Proving Ground Command during my entire five years at Eglin AFB. I owed a lot to the senior officers that I flew with—most had WWII and Korean combat experience—and I learned a lot from listening to them on many cross-country flights.

The Commander of the Air Proving Ground Command was Major General Robert W. Burns, whom I had the privilege of checking out in the T-33. He was an experienced fighter pilot but had never been checked out in jet aircraft. Within three flights I was just along for the ride, as he truly enjoyed his newfound toy. He would come down after work and really put the T-33 through a workout, probably the only joy that the General had all day. The pressure must have been very high because he finally had to dissolve his own command within a year. The year was 1957, the wars were long over, and the defense budget could no longer afford the Air Proving Ground Command.

I knew I was in trouble. My primary combat skill was as an F-86 fighter pilot and my normal home command was the Tactical Air Command, which was also being downsized as a lower priority than the Strategic Air Command. My advisor in Personnel informed me that my chances of finding any flying job were slim to none and I was probably going to be assigned as a scope dope (Radar Controller) in the Air Defense Command on top of some mountain in the far north looking out for a Soviet bomber attack over the polar route. The century series aircraft (F-100, F101, F102, and F-106) were coming into the inventory and my beloved F-86 was phasing out. I applied for Test Pilot school but did not have an engineering degree, so I applied to the Air Force Institute of Technology to return to college to obtain an Aeronautical Engineering degree. I was approved even though I only had two years of liberal arts but including Chemistry, Physics and math up through Differential Equations. I was assigned to Oklahoma State University in Still water, Oklahoma, for two years to obtain a BS in Electrical Engineering—not a likely opportunity for Test Pilot but better than a scope dope. My boss, Major O. B. Carter, who had molded me into a passable Air Operations officer, wished me well but he also had tried to help me find a follow-on flying job. It would be a long time before I would return to the cockpit in a real flying job, and I would fly as a behind-the-line pilot a little more than 100 hours per year for the next 10 years. Major Carter had driven me hard, but I owed him a big debt because he had greatly increased the breadth and depth of my flying experience and taught me air operations in the only outfit in the USAF that was operating most of the aircraft currently flying in the USAF. Our Paths would cross again. I left feeling that I was well ahead of my peer group but going back to college to catch up with them on education. Most USAF officers had college degrees.

Taking Care of Business

Sabra and I had been trying to have a baby ever since we were married. We both wanted many children and that was a big factor in our marriage. I had my semen tested for virility; Sabra had been examined by several doctors; and there was no reason that they could find to prevent conception. She had experienced female problems early in our marriage but those had cleared up. We applied to Catholic charities, the county, and finally the international adoption agencies, but after filling out many questionnaires and suffering through several home visits, we were deemed ineligible because we were not the same religion and I was employed in a risky occupation. I developed a strong dislike of social workers playing God. So now we were off to college again in Stillwater, Oklahoma, still without children.

We had sold our home in Fort Walton Beach for $9,600 and bought another home for $16,500 three blocks from the Oklahoma State University campus. Our neighbors were also USAF—Bob and Edna Hudgeons—and he was also a Korean Veteran and a fighter pilot. They had a brand new baby girl, Cindy, and they shared her with us. Bob and I had to study a lot because we had been out of school a long time and only had two years to earn our Engineering degrees. Most of our spare time was spent playing with Cindy and drinking coffee together. Bob and Edna turned out to be lifelong friends.

I was very worried because I had to carry a heavy course load of all engineering courses and go to summer school to obtain my degree within two years, competing with younger students who had been continuously in school for their whole lives. I also had to satisfy my minimum flying and ground school requirements, flying T-33s out of Vance AFB two hours away. My first two years of college I had spent chasing girls and drinking beer, so my grade point average was only slightly above average. Now with an adequate salary, happily married and motivated, I attacked my studies with vigor. After the first semester I felt good. I made the honor roll and my study habits were very good. All of the USAF students studied harder than the civilians and made most of the high grades. I enjoyed the entire experience and graduated with a lot of A's and only a few B's. I really enjoyed the learning experience.

Cindy Hudgeons, the puppy, and me

My study partner, Captain Les Daggett and I decided to invest in orange groves in Florida as a result of many cross-country trips together to satisfy our minimum flight requirements, and we ended up buying two young five-acre groves in central Florida. We contracted the maintenance with the intention of keeping the groves five years until they were producing a break-even crop and then trading them in for a 20-acre grove with the idea of accumulating about 80 acres of bearing groves before we retired. We had done our homework well but never envisioned the political trend of radical environmentalism that eventually increased our costs and liability to the point that the risks exceeded the ever-decreasing benefits. Walt Disney saved us by buying up a huge plot of land near us to build Disneyland, and we sold ours after 10 years at a modest profit.

Six months before graduation Sabra's father called. He had talked to Father Cunningham, the Fort Walton Beach priest, who allowed us to adopt a baby girl from an unwed mother in his parish. He picked up the baby, and Sabra flew home and joyously took the baby. We obtained a lawyer and he arranged the adoption and we were the proud parents of our first child, Mary Kellie Kennedy. Sabra brought Kellie back to Stillwater, and after six years we finally had our one-month-old baby. We were very, very happy. Then we were informed by the International Social Agency that we had been approved to adopt a one-month old Italian baby. We were very happy again—we had more than enough love for twins.

When they found out that we had just adopted another girl the same age they refused to let us have the Italian baby because it was not "natural." My contempt for social agencies knew no bounds.

We drove to St Louis for a weekend to show Kellie to my family. My grandmother, Momo, and my mother monopolized the baby. My family loved Sabra and the baby and knew we would soon be transferred to our new assignment, so we made the most of that weekend. My grandmother had always given me unconditional love, and she was now 91. We were very glad we made that last weekend trip, because that turned out to be the last time we saw her.

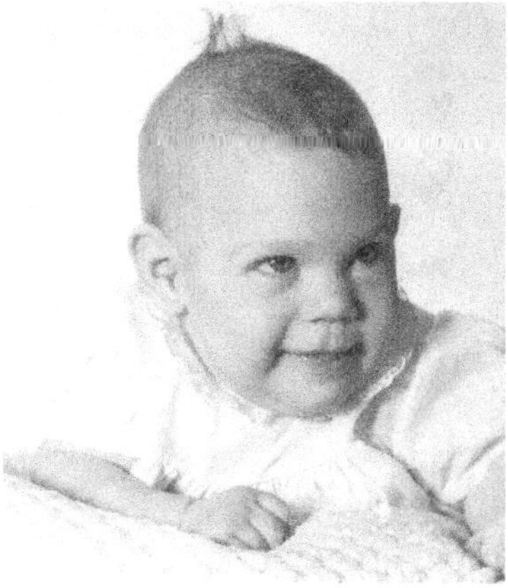
Mary Kellie Kennedy

My orders arrived two weeks before graduation, but I had to report for my new job within 30 days. I was assigned to HQ USAF Europe in DCS/Communications, Ramstein Air Base Germany. That meant selling my house, graduating, driving Sabra and Kellie to Sabra's home in Pensacola, Florida; then driving my car up to Newark, New Jersey, to ship our car; and then fly military air to Germany. That gave me 15 days after graduation to report to Germany. Sabra and Kellie would fly over 30 days later. I requested more time, but my new organization would not allow it. I tried my best to get everything done but could not make it. I drove Sabra and Kellie home to Pensacola, straight through overnight. The next morning I sold our new Ford Convertible for cash to a used-car dealer for a rip-off price and bought a commercial flight leaving that afternoon to New Jersey to catch my assigned military flight to Europe. I arrived at my new duty station on time in a very poor state of mind and reported to the executive officer who informed me that they did not have a specific job for me but would have one within a week. He suggested I use the time to look for housing in Kaiserslautern as there was no government housing available with a one-year waiting list. That was my introduction to the Communications career field. I had traded in our beloved convertible at a cash loss to make my reporting date because they had refused my request for an additional 15-day extension, and they did not even have a job for me! Our European overseas tour was off to a bad start.

CHAPTER 11

Europe

I was billeted in the visiting officers quarters at Ramstein AB until my family would arrive in 30 days. The next day I checked into the base and flight operations and then went into Kaiserslautern and bought an Opel automobile. I had been given a list of rentals by base housing and I'd had three years of German in high school and college; so I started my search for our home in downtown Kaiserslautern, Germany. Within two days I learned that I spoke high German and this was Bavaria and they spoke with a very different accent. The class system was still evident, because when they heard my accent they became very formal but also made it clear that we would be welcome. Many of them spoke English better than I spoke German, but communications were easy. I rented a two-bedroom apartment on the fourth floor of a six-story building. We would be the only Americans in the building. Only our landlord spoke a little English so we were to enjoy an interesting social life.

Sabra, Kellie and I in Ramstein

I had been promoted to Captain but could not pin on my rank for several months. I was in the Headquarters for all USAF units assigned to Europe, Africa and the Middle East in Communications, where the most experienced officers were handpicked for the Headquarters staff jobs. I was a fully qualified air operations officer and fighter pilot who had been issued to the communicators because I had a brand new BSEE degree in electrical engineering but no experience in Communications. Nobody wanted me. Fortunately, Lt. Colonel Andrew Wiegel, a pilot, hired me as his staff officer for Air Traffic Control Communications and Electronics, where I had experience in using these services for seven years. I was the only pilot among my peers in the Communications career field and thus envied because I received flight pay and took home $100 more every month. Fortunately my boss and his boss were also pilots so the envy had to be concealed.

Most of the communicators were non-rated, very sharp, and were experts in their chosen profession, which was necessary in our international National Allied Treaty Organization (NATO). We had to make certain that we could communicate and share targeting systems with our treaty partners in Europe, Africa and the Middle East in peacetime and wartime. Our NATO Air Forces and Ground Forces were numerically greatly outnumbered by the Warsaw Pact forces under the Control of the USSR. If they chose to start a conventional war with us in the European Theatre it was clear to me that we would have to go nuclear within a very short time period to prevent being overrun. Thus we were just an extension of our worldwide strategy of Mutually Assured Destruction and our presence in Europe merely placed our NATO allies under our nuclear protected umbrella. Most of our NATO allies spent a very small portion of their GDP on their military; and the US, United Kingdom and France with their conventional and nuclear-capable forces provided the bulk of our forces.

Sabra and Kellie arrived 30 days after I arrived. I had our apartment all ready but had not yet spent a night there because I was billeted in the Bachelor Officers Quarters awaiting my family. I met them at Frankfort Military Air terminal. They had been traveling for more than 24 hours. This was before we had jet airliners, so they had changed planes two times and had four refueling stops traversing the North Atlantic route via Greenland and the Azores. I drove them to have dinner at our Ramstein AB Officers Club and then to downtown Kaiserslautern to our very German apartment. We bedded Kellie down in a padded dresser drawer, pending arrival of

a baby bed and our belongings. Within thirty minutes of our arrival we heard the loudest church bells I have ever heard in my life. I had purposely selected an apartment just a few houses away from a Catholic Church but had blissfully ignored the very loud church bells. On the first GONG, Kellie's body jumped and she was totally awakened from a very sound sleep. Sabra was exhausted and put a pillow over her head, and I held Kellie for a couple of hours until she slept through the bells. Within one or two days we adapted. The bells tolled once on the half hour, plus enough times to broadcast the hour 24 hours per day. The human body can adapt to anything. The only thing that mattered was that we were all together under one roof. I felt at home again.

Sabra did not speak German, and all of the other wives in the apartment, other than the landlady, did not speak English. Nevertheless, they were very curious about us and they would gather for coffee in the morning and chat. A little German, a little American, and a lot of sign language and pretty soon Sabra was one of the gang. German babies are big and plump and Kellie was small and delicate and all of the German ladies that we met would ask for Kellie, cradle her on their robust breasts, exclaim, Ach so kleine! (Oh so little) and start stuffing fattening food down her throat. Poor Kellie could hardly wait to get back to her mother.

Assignments

Lt/Colonel Weigel assigned me staff surveillance for USAF Navigation Aids and Air Traffic Control Communications and Radars within Europe, Africa, and the Middle East. This included deployment of our mobile communications, radio navigation beacons and Ground Controlled Approach Radars. We had a Squadron of Air Traffic Controllers, Communicators and very talented com/electronic maintenance personnel equipped with mobile equipment, which could be airlifted to establish adequate communications and terminal all-weather navigation and approach control to any airfield within our area of operations and be set up and operating within a few days after arrival. Additionally, I was assigned staff surveillance for our USAF special com/electronic facilities in the Berlin area, a 25-mile diameter circle controlled by a joint U.S., British, French, and Soviet Authority per the WWII agreement. This area was deep within East Germany joined to the US, British and French Occupation zones by three air corridors. Within the Berlin Control area the air traffic was controlled and staffed by personnel from all four governments.

Each country operated its own airfield within the Berlin terminal area. We operated and maintained the area Radar Surveillance facility to direct all air traffic into and out of the Berlin corridors and terminal areas; but each country would control its own aircraft during terminal approach and climb out from its own airfield. The cooperation was good because everyone's safety depended on maintaining positive control and separation between arriving and departing aircraft. Obviously none of the aircraft were combat aircraft but each country had intelligence officers on staff, and there were probably more spies per square foot in Berlin than anywhere else in Europe. We obviously loved having the Surveillance Radar with its long range deep in the heart of East Germany and there were many tall antennae towers operated by each country. Each country was allowed to have its own security on their equipment to protect military designs. I requested, and was approved for, a Top Secret clearance as a result of this particular assignment. This clearance later gave me access to many other classified programs, which increased my value to the USAF; so I had lucked out again. Only Lt/Col. Weigel would take me, but he knew that my BSEE degree in Com/Electronics would make me more valuable in this job than my lack of communications experience. I felt very honored. Maybe the USAF personnel system was not as dumb as I thought. I had been given a real break. Now I had to deliver.

The first problem in my area was that our F-100 aircraft were equipped only with a low-frequency radio compass for enroute and terminal navigation. These receivers would be tuned to the frequency of an enroute or terminal radio station, and the radio compass needle would point toward that station. Each station had a unique three-letter identifier and would periodically broadcast their identifier in Morse code. The pilots would plan their routes from station to station to their final destination station and then make a published letdown and approach on the terminal station. The weather in Europe was much worse than the US. The weather statistics for the best USAF base in Europe were much worse than the worst USAF base in the United States. That meant that our pilots spent a lot more time flying in bad weather with only the obsolete low-frequency radio compass as a navigational aid in our supersonic single-seat century series combat aircraft. Unfortunately the low-frequency radio spectrum was very crowded in Europe with many stations very close together, so it became very important to identify each station correctly as soon as it was tuned in. We lost an F-100 in the Adriatic Sea because we believed that the pilot had tuned in the frequency of a beacon in Yugoslavia instead of the beacon in Aviano AB Italy. The frequencies were very close and could easily be incorrectly dialed in but not identified. One of our C-130 cargo flights to Trabzon, Turkey, strayed into the USSR territory during an approach to Trabzon. A Soviet Mig shot it down just across the border. In this case we believed the Soviets had deliberately set up a beacon very near in frequency and more powerful to draw the pilot into USSR territory. We could never prove our case but several incidents continued to occur. I wrote an article in our theatre Flying Safety monthly publication describing the technical problem in pilot language. That article became mandatory reading for all theatre pilots.

Our accident and incident rates involving weather penetrations were high, and our flying safety communities constantly complained that we needed voice recorder/reproducers in all of our air traffic control towers and ground-controlled approach facilities. The lack of this equipment greatly hindered accident and incident flying safety investigations. The Command had submitted a requirement for our Development Command to develop the equipment. It was approved but never funded. My old boss, Major O. B. Carter, was the staff officer in Operations responsible for Air traffic Control and he was constantly trying to obtain this equipment. He called me and asked if I could buy something locally. The Army had our electronic equipment procurement responsibility for some strange reason, and I called them. An Army civilian visited me and said he could put a specification out for bid on the equipment if I could show a valid requirement, a specification, and a fund citation. Major Carter already had a valid requirement signed by our General, and I enrolled help from our equipment technical representatives to prepare a technical specification based on our interface with our existing equipment. I contacted our focal point in our Development Center in Rome, New York, and told him we could satisfy our own requirement by purchasing the equipment from a local German vendor. No development was required. He had never been able to satisfy our requirement because of a lack of development funds, but he had procurement dollars and would provide them if we provided him with the data required by his organization. They were happy to get us off their back. One more hurdle was required. I needed to coordinate the installation of these recorders with the agency that would operate and maintain them, and devise a satisfactory maintenance concept. I worked hard on a briefing to present to the Colonel and his staff in charge of maintenance for all of the Air Traffic Control equipment in Europe, Africa and the Middle East. Colonel Fitzgerald was a big Irishman well known for giving briefers a hard time. I entered his briefing room and he was seated at a large desk surrounded by his staff of maintenance experts. There was a small table with a viewgraph projector in front of a screen to display my briefing slides. He promptly introduced me and the purpose of the briefing and told me to proceed. I dropped several of my slides on the floor, and when I bent over to pick them up, ripped the seat of my pants. Colonel Fitzgerald seemed pleased and immediately took charge. He called his secretary ori the squawk box and asked for a sewing kit to be sent in and then instructed me to take off my pants. After the secretary left, I took off my pants and gave them to Colonel Fitzgerald. I was in my Class A Blue Jacket, shirt, tie, with my long skinny legs, black socks and polished black shoes. He announced to my audience that Captain Kennedy had ripped his britches but that he (Colonel Fitzgerald) was going to sew them up while I presented my briefing. At first it was difficult to brief this audience of maintenance experts in one half of a Class A uniform, but I soon was chirping away like a fully clothed officer. We had a spirited discussion, and Colonel Fitzgerald and his staff gave me excellent advice on how to patch up my maintenance concept while the good Colonel sewed up my pants. The end result was a success. We all had a good time, and I recovered my dignity shortly thereafter. One year later the German company, Wolfgang Assman, delivered and installed these recorders in all of our Air Traffic Control facilities in Europe, Africa and the Middle East.

When I retired from the USAF 20 years later, the equipment was still operating reliably.

Good News

My old boss, Major O. B. Carter, had located a German Lady within our USAF organization who assisted military couples to adopt German orphans and we applied shortly after we arrived. Within six months we were notified they had a nine-month-old baby boy in an orphanage in Giessen, Germany, whom we could adopt. We rushed up to Giessen and picked up our second child.

KC and Kellie

He was half German and half American and was in a room with 25 other babies and one nurse. He was underweight, and his head was flat because he had been on his back most of his life. We rushed him to the USAF hospital and they gave him a physical examination. He was undernourished, but he passed the exam. We took him home. We named him Kevin Carter Kennedy after Major O.B. Carter, who had helped us find the German Lady and the German lawyer to handle the legal work. Kevin needed a lot of affection and holding because he had not had adequate attention. Sabra reshaped his head by keeping him from sleeping on his back and Kellie became his little momma and loved to cuddle him. We had to carefully build up his diet because he could not take our better baby food; however, inside of a year he was up to an adequate weight.

Bad News

We both received very bad news that first year. My grandmother died; and, within a month, Sabra's father died of a heart attack at age 57. We could not get back for the funerals, and we grieved together through this very difficult period.

After a year in our German apartment, we finally moved into USAF officers quarters on Ramstein AB Germany. They were four-story apartment buildings with two stairwells and eight apartments per stairwell. Sabra was losing weight and I could tell that she was not feeling well. She went to the USAF hospital where they had a German female doctor. She went several times over a period of a week or so, and finally the German doctor told her there was nothing wrong. Sabra lost consciousness outside the doctor's office; and, over the objections of the doctors and nurses, I took her medical records and helped her out to the car. I drove her up the hill to the US Army hospital emergency entrance. They took her vitals and screened her medical records. (The USAF German doctor had not performed any tests.) They drew her blood but there was no time to obtain results. She was hemorrhaging by now and her pulse was very weak. The US Army Doctor asked me to sign an authorization to perform exploratory surgery because she was sinking fast. I signed, and they took her off to the operating room. Within an hour after our arrival, she was in surgery. They found she had a tubal pregnancy; it had ruptured, and she had been bleeding internally for an excessive amount of time. They saved her life, and if I had not taken her out of the USAF hospital she would have died. She stayed in the Army hospital for more than a week. Major Carter and his wife,

Norma, loaned me their German maid, Erica, to stay with our children in my apartment until Sabra came home and got back on her feet. Erica stayed with her until she was well. We had an Air Force family now, and I think that is when I decided to make the USAF a career. We were friends with the Garters for the rest of our lives, and O.B. and I still see each other in our golden years.

I was furious at the negligence of the female German doctor and reported the incident to the Commander of the USAF hospital. The US Army doctor also provided supporting documents. He was as upset as I was, when he found that I had just seen the USAF doctor before I brought her to his emergency room. I did not have to wait long for action. The same female German doctor had once again told a wife her problem was in her head. The wife was admitted that same day for emergency surgery. This time it was the Colonel's wife, and the female German doctor was gone.

I had the best medical attention possible in the USAF but I never again felt comfortable about the care that my wife and children received, particularly in Germany. They were weak in female medicine. I believe Sabra had a female problem for a long time and the USAF never had the female expertise to diagnose why she could not have children.

Good News Again

Frau Dommermuth, the German lady who arranged our adoption, called us one day and asked if we still wanted more children and we said absolutely, Yes! She was impressed with our care of Kevin and his progress, so she offered us a four-month-old baby girl, who was in a private nursery near Frankfort. Her father was a married German industrial leader, her mother his mistress. We went to pick her up and she was in a very luxurious home with her own private nurse. The nurse was young and cried when we took our baby away. She had cared for the baby since birth and had become very attached. This was the exact opposite of Kevin, one of 25 babies sharing one nurse. The baby's birth mother made one request through Frau Dommermuth—to name her Martina. So Katherine Martina

Kellie, Kevin, Tina

Kennedy entered our family. Mary Kellie was a full-blooded American; Kevin Carter half German, half American; and Katherine Martina a full-blooded German. This was to be an unintended experiment on heredity versus environment. Sabra and I were in heaven. After 10 years of a very happy marriage, we were blessed with three of our own children.

Discovering Roots

I flew two or three times a month to maintain my required flying proficiency and was using the time to visit the European countries. On one of my flights to London I found a phone book for Northern Ireland where my father had been born and where he had been headed when he deserted us. His two brothers, whom I had found in the United States, Jim and Bill, both had said that they never saw him again in the US or Ireland. I looked under Kennedy and there was a Harry L. Kennedy on 52 Brookville Road in Belfast, Ireland. The address was familiar, but I do not know why. When I returned to Germany, I wrote a letter to Harry L. Kennedy asking him if he was my father and providing him sufficient information to refresh his memory. The letter was picked up by his wife and she was amazed to see that the handwriting was identical to her husband's and that the return address revealed it was from a Captain Robert W. Kennedy, whom she did not know. So she opened it and found out where her husband had been when he left her for ten years only to return to her eventually. She also learned that she was not his only wife. She gave the letter to her oldest son, Sydney, to answer. They both knew it was true because of the remarkable similarity of the handwriting. She did not show it to Harry and never told her daughter. It was Sydney's problem.

Sydney answered my letter and acknowledged that he accepted my letter as truthful. He briefly reviewed the circumstances within their family. He remembered that when he was very young, the family moved to America; and, in about 1922 his mother, sister Pamela, and himself returned to Belfast in Ireland without his father. His mother ran a linen shop for a living to support the family; her father was a wealthy and highly respected man in their community. My father, Harry, reappeared in Belfast about 1932. He had come home on a merchant ship, apparently broke. Harry had a hard time finding and keeping a job, and never again achieved the level of work that he had performed in America. They had another child after his return, but the baby died. Another baby, Jim, was born about 1938. Sydney explained in his letter that only his mother, Jim, and himself knew about my family. He invited me to come to Belfast and meet my two half brothers. I replied as soon as I could schedule one of my proficiency flights into Dublin, Ireland, and notified them of my estimated arrival time in Belfast.

I flew to Dublin and took a train to Belfast. They met me at the train station and we went to a pub in their neighborhood. We exchanged stories about our respective families and our own personal lives. Sydney was married, and they lived with his parents; Jim also lived with his parents. Sydney was about 43, I was 29, and Jim was about 22. We had a good time together and they were amazed that I looked so much like my father with many similar mannerisms. They must have had their mother's genes, because we did not look like brothers but had very similar feelings and experiences.

We got along famously, and finally they told me that our father would be walking home from his factory job and would come right by the pub. They wanted me to allow them to go out and bring him into the pub so we could meet each other. I thought about my mother who had never known if she was a widow (I never believed that she believed that he was dead although she always told me that he was). She had her life wasted. My sister, Doris, was deserted when she was six, a very sensitive time for a father/daughter relationship. She never recovered until she married her husband. I had never remembered him because I was a baby when he left. I did not want to meet him, since I knew I could not forgive him. So I declined the offer, saying it would be too disruptive to their family and it was best that we keep it as our secret. I had hoped to meet my half-sister, Pamela, but they did not make the offer. We continued our discussions as before and they took me back to the train station so that I could return to Dublin before the border closed. It was not safe to travel later because of the IRA terrorist threat.

During the train ride back to Dublin, I thought a lot about these events that had finally surfaced after years of uncertainty. I vowed that I would take care of all of the responsibilities that my father had assumed on his marriage vows. He was a bigamist, and we were his victims. I would do my best to make my mother's and sister's lives secure. Meanwhile, I had truly enjoyed my reunion with my half brothers and felt satisfied that I had finally unearthed the truth. I communicated by mail with Sydney and Jim for about 10 years, and then we tapered off to only occasional letters, until an unexpected resumption of relations with Jim in our twilight years.

I immediately related the total truth to my mother and sister by telephone upon my return to Ramstein AB Germany. They both welcomed the news, since it resolved our family's biggest uncertainty. Harry was alive and well living with his first family, and we were now defined. I am certain that all of our family was now at peace and our conscience was clear. My grandfather remained as my only male role model.

Air Safety

I had been flying for over seven years, had more than the required flying time, held the highest instrument flying rating, and was awarded my senior pilot wings, which consisted of placing a star on top of my wings. I felt more comfortable flying in the European weather and twice had to land my T-33 in weather below GCA minimums when a pea soup fog rolled in and all nearby airfields went below minimums. In my job I also learned details about the problems and technical capabilities of our Air Traffic Control Radars and Navaid equipment.

We had our latest fighter F-105 aircraft coming into the theatre to modernize our forces, and they were equipped with TACAN navigational equipment only. We did not have total TACAN theatre coverage, lacking complete enroute and terminal coverage in the eastern Mediterranean. The military used TACAN and civil aviation used VOR/DME. Where possible, we combined these two facilities into a single Navaid to serve both civil and military aircraft. The Italian, Greek and Turkish governments had very poor cooperation and had not worked out a frequency allocation plan to cover their portion of the eastern Mediterranean. The Turks and Greeks were constantly feuding over Cyprus, both governments had had recent military coups, and there was very little communication between their military and civilian government. None of their military aircraft had TACAN yet, and many Italian, Greek and Turkish commercial airliners were equipped with VOR and had an operational ground VOR environment. Italy, Greece and Turkey would not need VOR/DME until their commercial aircraft were equipped to receive DME sometime in the future. So they were in no hurry to make a TACAN/VOR/DME frequency plan for the Eastern Mediterranean. It was a USAF only problem, and our attempts to work out a plan via international frequency allocation and international airline organizations were to no avail. Frequency allocations were a national asset, and each country had to approve all frequency plans within their countries. The VOR/DME TACAN frequency plan covered their entire territories and had to be centrally planned to avoid adjacent channel interference and overlapping coverage on the same channel. We were at a standstill.

I obtained a very large piece of graph paper and a large compass and went to work using the planned location of all the civilian VOR stations and the desired locations for our USAF TACAN terminal locations. Where they could serve both military and civilian traffic, I used a combined VORTAC station. I considered all of the technical requirements and made a TACAN/VOR/DME frequency plan for the Eastern Mediterranean and North Africa areas. I gave it to the USAF Major in charge of coordinating our frequency requirements with the US representative to the International Frequency Allocation Coordinators. He had us prepare a technical briefing to make the three governments feel that all of their technical and safety needs had been covered; then the Major and the head of our US delegation presented it to the three nations at the next meeting of the Eastern Mediterranean governments. It worked. Since there were very few communications between these countries, they communicated freely with the US attendees, and the frequency plan was approved subject to a technical review by each country. I attended the meeting in a very plush hotel in Geneva, Switzerland, and was treated to a superb dinner that evening by the senior US delegate. He also invited my family to use his chalet in Zermott, Switzerland, for a week of our choice. It slept 14 and was high in the Swiss Alps, accessible only by cog train followed by a horse-driven sleigh. The picture window looked out directly to the Matterhorn.

Swiss Vacation

We took our whole family and our German maid to help with the children and drove to the cog station below Zermott. We enjoyed the steep trip on the cog train up to Zermott. The air was very thin and we had a hard time loading our luggage onto the horse-driven buggy. It was even more difficult climbing the last very steep path up to our chalet through the snow. Once settled, we relaxed, built a fire in the fireplace and gazed out over our Christmas-card view of the Matterhorn. The children got up early and we all went out and played in the snow. I went down and rented the ski equipment that we would need and Sabra and I ventured forth. We started on the lower slopes near the village and gained some confidence for the more challenging slopes. We had skied in Garmish, Berchtesgaden and the Zugspitz previously, but Zermatt was the championship course. We were the only Americans there and we met some British couples in the local inn where we sipped hot buttered rum in front of the fireplace. We invited them to our chalet and offered them scotch, which delighted them. I had no ice, which they did not use, but Sabra and I did. So I went outside and put snow into our scotch. Before the evening ended we had the Brits drinking scotch and snow. I doubt that it caught on in London.

The second day, Sabra fell and hurt her arm and ankle, so all of a sudden I was solo. Major Walt Payne called us and he could not make it so we had the chalet all to ourselves. By the end of the week I worked up my courage to go all the way to the top. You could ski down into Italy or Switzerland and I picked the trip back down to the inn near our chalet. The ski lift was full all the way to the top. I knew better than to try this challenge without a partner but foolishly did it anyway. We all got out of the lift and I bent over to fasten my ski boots and heard a rip. I felt the cold wind on my rear end and knew I had ripped my ski pants big time. I made my way to the back of the group planning to be the last to push off. There was a group of four people waiting for friends on the next lift so I knew I would have to slip by them quickly and push off. I did, but I had not looked first and planned my push-off. It was steeper than I had ever experienced and I was out of control and going faster than I ever had been. It seemed like forever before I felt I was in control and I did not feel comfortable until I was half way down. I finally had the inn in sight and I pulled up short of the inn to signal Sabra to come and help me cover my ripped pants. She finally spotted me but could not understand my problem. She finally came over and gave me her jacket to cover my bottom. We all had a wonderful time in Zermatt. The children, the maid and Sabra spent the time sledding down some nifty slopes. We reluctantly left Zermatt and drove to Geneva to check out Sabra's arm. She had a hairline fracture, which surprised us, but it was fortunate that she had discontinued skiing. The Swiss doctor put a cast on it and we drove home.

Moving Onward

We were notified shortly after we had moved into government quarters in Ramstein that our part of Headquarters United States Air Forces Europe was moving to Weisbaden, Germany, a more metropolitan city famous for its spas and gambling casino. We moved into nicer quarters and met the Lt. Col. Sculley family, a wonderful couple with four boys and a girl. We enjoyed each other' company and took many trips with all of our families. Our whole stairwell was a group of fun-loving people. The men had a Thursday night nickel-dime-quarter poker club and it was better than Shakespearian theatre. We had the youngest children and so they became spoiled by the adults and older kids in the stairwell. You could not ask for better neighbors.

This was my first time in a Combat Command in a Headquarters with a brand new Top Secret security clearance, which gave me a new visibility into political and military decisions at a higher level. President Kennedy was now in office and tensions were growing in Europe. We had been building a Medium Range Ballistic Missile site in Italy and Turkey, which could target the USSR and which really bristled the Soviets with their paranoia on homeland defense. Our role in the unsuccessful invasion in Cuba also put them on edge but displayed a fatal U S weakness when we removed all support for the Cuban invaders at the last moment. Sure enough, they built the Berlin wall to escalate the conflict and now our forces in Europe were on alert. All of a sudden my part-time Berlin project became full time and I departed for Berlin for an extended stay.

President Kennedy called up the Air National Guard and several Reserve units in the United States with the intention of deploying these short-range fighter units to Germany. We responded to the Pentagon that all of our Air Bases in Germany were completely saturated, as well as all of our Main Operating Bases in France. After our third time stating that there were no available bases within striking range of the enemy lines and totally inadequate supplies of munitions, we received a message from the Pentagon stating that this call-up of the reserve and guard forces was to awaken the American people; the President had stated these fighter wings were coming, and we had to make room for them. They flew these fighter wings across the Atlantic and landed at our secondary Disbursed Operating Bases in France, where they could not reach the front lines and without an adequate supply of weapons. Our Major General in charge of Operations for 17th Air Force went to each one of these bases to formally greet the commanders of each Wing when they arrived. When he greeted the commander of the fighter wing from Massachusetts and asked him to call him personally if he required assistance, the commander reportedly arrogantly replied, "If I need help I will call Jack" (Kennedy).

These civilians were called away from their civilian jobs to fly over to France to inadequate bases where they could not even enter the initial battle should one occur. They had to have figured out that they were not adding significant capability to our war-fighting capacity. A significant number of military officers who had access to this Top Secret data, now declassified, were beginning to lose confidence in this reckless President and his Ivy League advisors. Later the President, through Secretary of Defense McNamera, ordered a battalion of armed US Army troops to convoy up the main corridor from West Germany through East Germany into Berlin. This was allowed under the provisions of the WWII peace agreement but was certainly a needless provocation during these very tense times. At the last moment he directed that none of the weapons would have ammo loaded but be available. The President and his advisors did not trust their military again. They had the CIA plan the Cuban invasion instead of the military and only coordinated their very poor invasion plan with the Pentagon at the last moment. Thus this battalion journeyed up the highway through the corridor to Berlin deep within East Germany with (allegedly) unloaded weapons. By now our President had a credibility crisis with his military, and that was very dangerous because he was our Commander-in-Chief. I felt very uneasy because he was displaying weakness, which an enemy can easily exploit. We would always be loyal to our President, but our confidence in him was being tested. I only discussed these matters with my trusted friends but I could tell our senior officers were troubled. It would get worse.

Meanwhile I was in Berlin, busy as a bee. Secretary of Defense McNamara had decided that the Soviets might exploit electronic jamming to prevent our aircraft from flying into and out of Berlin. He had authorized emergency funding to install all of our secret electronic countermeasures from our Air Defense systems protecting our homeland, as well as those in development on our communications and radars within Berlin. I had staff surveillance over all of these, so my knowledge was needed onsite to expedite implementation. When this plan first surfaced, I had pointed out to our four-star General that we were about to install all of our very secret electronic countermeasures, planned for our Air Defense system protecting the entire United States, into our Berlin facilities within a 25-mile diameter area deep inside enemy territory, which contained a very large number of spies and enemy listening and recording capabilities. A compromise was very probable. The General forwarded these concerns to the Pentagon to no avail. The funds and many electronic contractors were soon on site. The project "Bamboo Tree" was underway and a full colonel replaced me. He kept me onsite until he felt comfortable, and I returned to my bride and our children in Weisbaden. Once again I was in the right place at the right time and received very favorable visibility.

Sabra and all of our wives had been thoroughly briefed on evacuation routes to a location in France where they would be safer if war should break out. They had our emergency rations, extra gas cans and water in the car. She and the children were very happy that I was back and I loved coming home. Separation was miserable but homecoming was heartwarming. I had extended our three-year tour into four years to give me a better chance at a flying job in fighters but my chances did not look good. Now the communicators who did not know what to do with me when I arrived decided they wanted me in communications. I was still a reserve officer and the promotion opportunities were limited above the rank of major. I anticipated a problem in making major because I had no home command, one short tour in Tactical Air, one long tour in the Air Proving Ground Command (now dissolved) and now one overseas tour in Europe as a communicator. Three different Commands, three different specialty codes, and a reserve Captain. I had done well in Squadron Officer School and had earned my BSEE with honors, but I was not competitive with my peers, particularly those in the combat commands who had continuous progression within their command and were well connected. Several of my classmates who were assigned to the Strategic Air Commands were already Majors. The services allow accelerated promotions to deserving officers. They deserved it because many had been waiting to take off on missions deep in Soviet territory with a very low probability of return in a nuclear war. I had never reviewed my fitness reports from my bosses but felt that I would rate below my peers because of a lack of career focus and excessive time in non-combat-ready flying positions. I had to increase my value to the USAF and once again I applied for an MSEE when we returned from Europe.

Lt/Col Weigel had rotated back to the United States, and I had a new boss. He was a non-rated Major who had been a long time communicator, and he was not comfortable with the Air Traffic Control Communications/Electronics. Our whole office worked hard to get him up to speed but it was difficult. He always hesitated about approving our messages and letters and the slowdown in our office became very noticeable. He had a very hard time with my Berlin electronics and I tried very hard to get him up to speed but to no avail. He looked bad to his boss and I looked bad to him. It was a lose-lose deal. My last six months were not happy at work.

We had had a wonderful time in Europe touring Germany, Holland, Belgium, France (Paris), Spain, Italy and Switzerland. We took our children

everywhere except Paris and after two days we missed them so much we came home early. We made many lifelong friends and became even closer with Major O. B. Carter and his wife, Norma. We received our assignment to the Air Force Institute Of Technology at Wright Patterson AFB Ohio to study for two years to obtain a MSEE degree specializing in Guidance and Control. Presidents Kennedy and Eisenhower had debated during the election whether or not the Soviets missiles were more effective than ours, and I suspected that would be the reason that the USAF was breaking out Guidance and Control and teaching it in their own accredited Air Force Institute of Technology. Thus it was back to school again.

All of the equipment was installed in Berlin before I left; and, sure enough, during checkout they inadvertently had radiated in a secret mode that compromised that mode to the Soviets, negating its effectiveness later. The Soviets probably had detected and recorded the signal and worked out a countermeasure. The Medium Range Ballistic Missile installations in Italy and Turkey were soon to be operational and the Soviets were not happy. It was a tense time, but war had been avoided during our tour. Now we headed back home once again to study for an advanced degree.

CHAPTER 12

Stateside Again

Our family flew from Frankfort, Germany, to Idlewild Airport, New York (now JFK International Airport), arriving four hours late and missing our connection to New Orleans. This was a dramatic change from our flight to Europe. We flew on a commercial jet aircraft, which was a single six-hour flight instead of the very long 15 hours with three stops for refueling that we both had experienced on our flights to Europe. That was the good news. The bad news was that it left Frankfort four hours late. I had bought commercial tickets for all of us to fly from New York to New Orleans, and now we had to re-book on a later flight. It cost us a lot more money because there were not enough seats to accommodate all five of us. We had to upgrade one of us to first class to get on that plane or spend the night in New York and catch the early morning flight. Since the children had been awake for 15 hours and had three hours more to go, we split up, and Sabra and Tina sat in first class to New Orleans.

I had shipped my little Opel to New Orleans 45 days earlier, so when we landed we picked up our luggage and I took a cab to the Port of New Orleans. Sabra and the exhausted children then waited for me in the terminal to pick them up and drive the 200 miles to Pensacola for a joyous homecoming with her family. It was very late when we arrived, and we all kissed Grandma Sanchez and collapsed into bed.

We could have slept much longer, but all of Sabra's sisters and her brother had arrived right after breakfast and could not wait to see the children. They were really spoiled during our visit, as her sisters played with our children and we marveled at how grown up our little sisters had become. Cathy, the youngest, was already seventeen.

We drove up to St. Louis and stayed with my mom and Aunt Gladys. After my grandmother died, Mom and Glad rented the second floor of a duplex in Clayton. Our old neighborhood was badly deteriorated with a very high crime rate. We drove through it, and it looked like a war zone. It was very sad to see. The entire west end of St. Louis was looking very rough.

Mom and five of her six grandchildren

The move to the suburbs had been a significant struggle with severe financial problems. Property values and tax revenue had plummeted. Our old neighborhood was a slum.

We had a nice visit with Mom and Glad, then drove up to Dayton and stayed in government quarters until we bought a three-bedroom, two-bath home in Huber Heights, one of many suburbs that had sprung up after WWII. We moved in as soon as our furniture arrived from Germany. School started shortly thereafter and it was very challenging. My mother and Doris and Phil came and visited us before school started, and Mom finally had the opportunity to play with her grandchildren.

The Air Force Institute of Technology: 1962-1964

There were 20 of us in my Graduate Guidance and Control Class. It was a new specialty in education, an outgrowth of the space and missile age. We used textbooks from MIT, Stanford and Colgate for the new disciplines of astronautics, guidance and control, and advanced flight controls and the standard texts for our math, physics, and electrical engineering courses. Eighteen of my 20 classmates were brand new second lieutenants who had just received their BSEE degrees from leading universities, and we were graded on the curve. For two years I did nothing but study— sleeping, eating and flying only as required.

Sabra and my children saw even less of me than they did in Europe. It was very hard, and I could never have made it without a very understanding wife and mother.

We were required to prepare an original thesis and it was a significant contributor to our final grade. I designed an autopilot for our Boeing 707, which manually flew zero-gravity parabolas to provide astronauts about 50 seconds of zero gravity in the passenger compartment, which was padded to reduce the return to normal G forces during pull-out from the parabola maneuver. I obtained the aircraft aerodynamic and autopilot data from Boeing and designed a program to generate the required maneuver in the autopilot. The design required a derivation of a matrix that represented the acceleration G forces at the center of gravity of the aircraft with respect to inertial space. Then I designed an acceleration autopilot to fly the parabola, such that these aircraft acceleration forces exactly canceled out the earth's gravitational forces at the center of gravity of the aircraft. Einstein's Principle of Equivalence established the validity of the equivalence of acceleration forces and gravitational forces. Thus my autopilot could simulate any gravitational environment at the center of gravity of this aircraft between zero (space) and one (earth). This would allow Moon and Mars gravity simulations. I then simulated the entire scenario on the computer and flew the parabolas attaching these parabolas and the gravity printouts as an attachment to my thesis. My professors were impressed.

Only 18 of us received our MSEE degree, with two receiving a certificate of completion. The Air Force had received accreditation for our graduate degrees one year before our class started, and they had applied very rigorous standards to keep their Air Force Institute of Technology on the leading edge of the new disciplines. All 19 of my classmates were assigned as astronautical engineers to the Ballistic Missile Division, Space Systems Division, or The Central Inertial Guidance Test Facility within the Air Force Systems Command to develop and test our emerging capabilities. I was the most senior graduate and I was assigned as an astronautical engineer in the Air Force Avionics Laboratory at Wright Patterson AFB Ohio. We did not even have to move!

While I was busy studying, the Cuban missile crisis unfolded, and I learned that President Kennedy had turned back the Soviet fleet accompanying the missiles to be installed in Cuba by canceling our medium range ballistic missiles in Turkey and Italy. It was a Top Secret program so it was not widely known and not declassified for more than 20 years. It occurred to me at the time that perhaps that was Khrushev's objective when he deployed his missiles to Cuba. It appeared to be a clear victory for Kennedy, but it looked like a draw to me: "If you remove your missiles in Italy and Turkey I will not install my missiles in Cuba." The American people never were aware of this compromise until it was revealed more than 20 years later, and people never made the connection between the two events. President Kennedy was assassinated before I graduated but not before he entered a new battlefield in Vietnam. In a few years I would be deeply involved.

I was in the primary zone for promotion to Major, but I still was a reserve Major and the regulars had a strong advantage. Nearly everybody is promoted up to the rank of Major; however, a significant number of Captains, particularly reserves, fail to make Major. It basically means that you will probably be able to stay in the service for 20 years and be able to retire with half of your basic pay. I had been in six different commands, never in the same one twice. I reviewed my ratings in San Antonio Texas. I had never seen them and had never been counseled on them, so it was a shock to read how all of my supervisors had rated me for the past eleven years. My flying ratings had very short narratives and omitted my excellent scores on aerial gunnery. Most of my accomplishments at Eglin AFB and in Europe were omitted, and my rating officer (the non-rated Major who was my boss for my last six months in Europe) rated me as average. That was the kiss of death, since officer ratings had been over-inflated for my entire career. Fortunately, the endorsing officer, Colonel Sam Huey, who had a Top Secret clearance, disagreed and gave me a superior rating with a full-page description of my accomplishments. That was my only chance for promotion, since the stark difference between my supervisor's rating and the endorser's rating was a guarantee that everyone on the promotion board would study that carefully. I returned home very discouraged but looking forward to my new career as an astronautical engineer in the Avionics Laboratory. A few months later my Laboratory Commander notified me that I had made Major in the REGULAR Air Force. I was no longer hired help. I was finally on the first team.

Just before I graduated, we were

notified that our two German children, Kevin and Tina, had been accepted as US citizens. We all went down to the Federal Court in downtown Dayton, and they were sworn in with many other immigrants in a nice ceremony by the judge. They both received flags and felt proud to finally be Americans. They had been on German passports for several years.

Space Defense: 1964-1966

I was assigned as the project manager and engineer of an exploratory development program for Space Defense and had a Research and Development budget of about $5,000,000 per year. The Space Defense missions included Satellite Interceptor, Rendezvous, Inspection and Ballistic Missile Interceptor. It was a highly classified area and only a small number of people within the Department of Defense were aware of the details involved. My previous clearances in Berlin allowed my special clearances required for this program to be approved rapidly. The priority mission was Satellite Interceptor.

The USAF operated a Satellite Interceptor capability on Johnson Island in the middle of the Pacific Ocean. The island was selected because it was furthest from the nearest populated land mass, Hawaii, which was 800 miles away. The Interceptor was a small nuclear warhead on top of a Thor missile. Our tracking stations throughout the world could find and track a newly launched soviet satellite and within a short time be able to predict its future location with sufficient accuracy to launch our interceptor at the predicted location. The predicted location encompassed a sphere of uncertainty that required a nuclear weapon to assure a kill throughout the entire sphere of uncertainty. The nuclear warhead would be detonated at the predicted impact point, and the lethal radius of the warhead was greater than the sphere of uncertainty of the target location. We had never detonated a nuclear weapon in space and our scientists, concerned with collateral damage, needed to know the effects. Thus they did a full test of this new capability and detonated a Thor-launched nuclear weapon about 100 miles above Johnson Island in the early 1960s. The effects were significant in Hawaii, and we learned much more about Electromagnetic Pulse (EMP) phenomena. It was not good news. The detonation was kept classified and a cover story was generated to explain the mysterious electronic failures in Hawaii. I am certain the Soviets and others had recognized the location and true nature of the detonation and increased their surveillance of all activities on Johnson Island. When I came on board in 1964, the emphasis was to replace the nuclear warhead on the Thor missile with a non-nuclear Terminal Homing Vehicle (THV) that would detect, track and lock-on to the enemy satellite with an appropriate sensor at a range sufficient to separate the THV from the Thor missile, ignite the Propulsion unit, and home into and destroy the satellite. My job was to select the optimum sensor and determine the design requirements for the propulsion, fusing, non-nuclear warhead and the THV structure. Thus our THV would replace the nuclear kill warhead with a terminal homing missile to achieve a much smaller miss distance, enabling the use of a conventional weapon to kill the enemy satellite. The closing velocities of the two vehicles were enormous, but it would be similar to hitting a bullet with another bullet. The risk was very high, the technology was not yet developed, and that is why this task was in an exploratory development program. (This is the initial development step just after basic research.) Once again I had been well assigned. Most of the tasks were like doing the homework problems in my recently completed MSEE course, and the space environment was more linear than working similar problems in the atmosphere.

My boss was a civilian Physicist who was very quiet, but he taught me more about the scientific method and a systematic way to pursue problems, always questioning and probing for truth. He stressed listening and analysis, always seeking a finer resolution of details. Jerry Pasek was the brightest scientist I have ever known. He was very experienced in aircraft avionics but, like most of us, he was new to space. Jerry was an avid researcher who kept abreast of the latest developments. It was exciting to work in areas where there were no textbooks, just the most recent contracted studies and technical reports.

We started the effort by funding three of the leading space defense contractors to independently do the preliminary system tradeoffs and design studies. They would bound the performance requirements for the individual THV subsystems (sensor, guidance, fuse and warhead) as dictated by the known details about our Thor booster missile, the Johnson Island facilities, and the accuracy of our missile tracking system (which had already been done for the nuclear warhead version). Each would deliver a simulation program capable of testing their design and others with the ability to try out and size various THV subsystems for a variety of intercepts.

I started working on the most critical subsystem—the sensor to detect, lock-on and track the enemy satellite to guide our THV to the target. There were three candidates, RADAR, Infrared, and TV. TV was the lowest risk that could detect the satellite at an adequate distance due to the reflected sunlight, except when the intercept was in the earth's shadow. Unfortunately, this was an unacceptable operational constraint, so it was the least desirable. RADAR was the heaviest sensor but provided range as well as angle data. Long Wavelength Infrared sensors had not been developed or even tested but appeared to have many advantages. We were looking for a satellite at earth temperature against a space background, which was four degrees above absolute zero. That provides a very high signal-to-noise ratio that would allow us to detect a block of ice at distances of more than 100 miles. It would take more time and higher risk to develop and test the infrared sensors but the payoff was higher. We decided to pursue all three sensors and keep our eyes open for others.

I started the feasibility study for RADAR. My concern was that when a RADAR signal is reflected off of a target it comes back in lobes that have very high peaks every few degrees, and very low nulls between these peaks. At very high closing velocities, if the initial geometry was in a null, the satellite would not be detected in time for the intercept to occur. I had three satellite targets made and took RADAR reflection measurements at many different frequencies for every angle of approach. The measurements were in analog format because all of our RADARs were analog. I needed the data in digital format so we could make mathematical manipulations on the data. We digitized all of this massive amount of data and began our analysis. The nulls negated the RADAR solution, even after we tried to average them out by changing frequency for each set of pulses. It increased the average signal but not enough. My experiment was a success in a negative way, because it eliminated one approach but was very successful in that it found a processing technique for solving the most difficult airborne RADAR problem of detecting a low-flying enemy aircraft hiding in the RADAR ground clutter. Jerry Pasek was delighted that for the first time he had access to RADAR data in digital format and he could try out signal processing schemes to extract more information. He had me process my data using the Fourier frequency transform, and all of a sudden he saw how it could be used to filter out ground clutter. He instrumented it in our RADAR test bed installed in a B-25 flying out of L.A. International and all of the ground clutter disappeared. It was a major breakthrough in airborne RADAR and has been used in all of our aircraft RADARs ever since. Similarly, another division used digitized RADAR data to design the early stealth aircraft. So, inadvertently, I had developed a process that became very valuable in fixture RADARS. Today they all are digital. I wrote a technical report on the process and the measurements, The Spatial and Spectral Characteristics of Two Satellite Targets. It was published, and I received the Avionics Laboratory Kulpa Award for the best technical report of the year. I was very honored, because fewer than three percent of the Labs Scientists and Engineers were military, and a civilian usually won the award. Parts of my report were included in the next edition to *Skolniks RADAR Handbook.*

Now we concentrated on our Long Wavelength Infrared (LWIR) sensor. Scientists in another Government Lab had designed and built detectors in the LWIR frequency range that was optimum for our application. They were copper-doped germanium chips cooled to liquid hydrogen temperatures, and we were working on mercury-doped germanium cooled to liquid nitrogen temperatures, which would be much more supportable in our sensor. The copper-doped detectors were available now and we were offered a piggyback ride on a Thor Launch at Johnson Island. Anxious to demonstrate the feasibility of LWIR technology, we had our contractor, Hughes, design a very simple search track sensor with our LWIR detectors and a small supply of liquid hydrogen and the electronics and structure to interface with the Thor missile. Our team then journeyed to Johnson Island via Honolulu. We traveled commercial air and stayed in commercial hotels to conceal our military identity. We had to fly military air from Honolulu to Johnson Island because there were no commercial airlines that went there, and it was restricted to visitors who had the appropriate security clearances. There was only one USAF C124 flight per week to Johnson Island, and I am sure that the Soviets knew every name that traveled to that island after the nuclear airburst in space above the island. If I had failed to be listed as a person of interest to the Soviets due to my Berlin activities, I was certain to be listed now. This became a concern to me later.

We were on the Island for several weeks, making adjustments, and finally were ready for launch awaiting the launch of the target. We were not the primary pay load but an experimental payload, so we waited until the events required by the primary payload were satisfied. The launch decision was finally made and we watched as the launch team completed their countdown and launched the Thor, which went off with a mighty roar. It was very impressive.

The telemetry data indicated that the primary payload had been successful, but the data from our payload indicated we had detected many objects; however, much more analysis would be required. We left Johnson Island on the next flight to Hickam AFB Hawaii and then flew to our respective homes in the U.S.

When the telemetry data had been processed, we began our analysis and concluded that we had detected some of the debris from the explosive squibs that had released the cover for the optical payloads and possibly detected some stars. The primary payload was successful, but we had failed to detect the target, probably because our closing velocity was so large that our scan rate had a low probability of putting our detector on target. The evidence was sufficient to conclude that we had more than sufficient sensitivity to perform the satellite interception mission. Now we had a candidate primary sensor with a TV sensor as backup. Sensor feasibility was established. Now we plugged our sensor data into our Satellite Interceptor mission simulations and could size our propulsion, fusing, and warhead requirements. We had developed and demonstrated the performance of LWIR sensors in space for the first time and many applications would soon follow.

The Space System Division decided we were ready to begin an engineering development for a Thor-launched Satellite Interceptor at Johnson Island, replacing the nuclear warhead with a sensor-guide Terminal Homing Vehicle, a missile with a conventional warhead on top of a missile. They issued a Request For Proposal to the three system contractors who had designed the THV system simulations for my exploratory development program. I was selected by the Source Selection Authority to head up the evaluation panel for the guidance and control subsystem. The program was still special access with very tight security controls, so when the three proposals were received, all of the evaluators were confined to work within a secured area and warned that any security breaches would be punished to the full extent of the law. I shared an apartment in El Segundo, California, with another evaluator and we wore civilian clothes to conceal our military identities. Sabra did not know where I was for months until we had completed our evaluation and presented our results to the Assistant Secretary of Defense for Research and Development in the Pentagon. I was one of the four briefers describing our evaluation of the competitive scores for the three contractors on the Guidance and Control. My briefing revealed some remaining risk in the sensor, fuse, and warhead tradeoffs with the fuse providing the greatest uncertainty. Our evaluation was accepted and Ling Tempco Vought became our prime contractor to develop our THV.

Sabra had not known where I was for several months and had found a four-bedroom home just three blocks away from our home in Huber Heights. Since we were going to stay in Dayton, Ohio, for another tour, she was looking

for a larger home. She found one that she loved. It was a brand new two-story, four-bedroom home in our same school district and the children loved it. She tried everything she knew but could not get in touch with me. The price was $ 20,500 and the builder allowed her to hold it for a $10 down payment! She called Jerry Pasek, my boss, and he arranged to pass the message to me. I was authorized a call home and was elated with her description of our new home. When I returned home, we signed the paperwork and moved into our new home.

Molly, Sabra's sister, had moved to Dayton and married Ben Hegman, who was a wonderful husband and father. They lived close to us in Huber Heights and we grew very close to them and their children. We could always rely on one another in times of stress.

We also had made some very good friends in our neighborhood, mostly military couples with children who played with our children. Ray and Mel Grazier and Chuck and Rose Lysaght were our favorites, and we spent many weekends together with both families and the children. That was the best time of our life—when the children were growing up and we had a lot of fun in the neighborhood. My favorite memory was on Saturday morning when all of our kids would come into our bedroom to cuddle and play games. We hiked a lot in the beautiful Ohio forests. Glen Helen, near Antioch College, was our favorite. We had a trail with a swinging vine, a rope

bridge across a creek, a rock rapids that we could ride down on the seat of our pants, rock towers to climb, and a nature museum. We would take our collie with us on our long hikes, and when I gave our kids the command to scatter, they would all run away in different directions and we would watch as the collie became excited, barking for attention and nipping at the children to herd them back to us. We normally took our children and their friends on these hikes. Sabra was a Blue Bird Leader and had a group of girls that she would take camping overnight. Most of us were military and were gone more than our civilian neighbors, but our wives handled it very well. All of our problems were little problems, and love enveloped our family and neighborhood. We always had neighborhood cook-outs, snowball fights, sledding, volleyball, or hide-and-seek games with both adults and kids. We preferred that all of the kids play at our house and they all stayed good.

Tina's 2nd birthday party

I continued to make periodic visits to Johnson Island, and most of my effort went into working on supporting technologies for our LWIR sensor. Industry had developed better detectors that could operate at liquid nitrogen temperatures, which were much easier to support. We also had to cryogenically cool the entire telescope to maintain the detector sensitivity. The fuse still remained our primary uncertainty. During my last visit to Johnson Island, President Lyndon Johnson announced that we had a Satellite Interceptor system in place. That had to be the nuclear warhead capability, but I do not remember that he specified it as such. This really surprised us, considering the rigid security environment. Many of us worried that it could be the start of space warfare. I believed that we had more to lose than the Soviets, since I believed that we were far ahead of the Soviets in exploiting space for precision surveillance, navigation/guidance and communications, which were certainly going to increase our surface war fighting capability in the fixture. Therefore, I did not relish the idea to start a space race with satellite interceptors.

Johnson Island was a big barren sandbar in the middle of the Pacific Ocean when our military dredged up more sand to make it an island and built a big runway on it during WWII. It was an ideal location for our Satellite Interceptor launch facility. We had a gorgeous barrier reef with crystal clear water just off shore, where we enjoyed snorkeling during our long countdowns for launch. When you first looked down through the clear water, you felt as though you would fall down onto the reef because the water was so clear. There was a plethora of sea life on the reef with every color in the rainbow present. The reef was forbidden because for some mysterious reason all of the fish on that reef were dangerously poisonous. We imported natives from Hawaii to work on the island but forbid them to fish, warning them that the sea life was poisonous. They had fished their whole life and these fish were identical to the fish in their native Hawaii. Several times a year one or two Hawaiians would spear a fish on the reef and fall fatally ill shortly after eating it. They would be air-evacuated back to Hawaii in critical condition. The other hazard was sharks. We saw them often in the vicinity of the reef while we were snorkeling. One day Jack Welch, the Vought THV program manager, the Hughes LWIR project manager, and I were out just a little bit further from the reef when we noticed a large hammerhead shark circling us. We all promptly surfaced and swam in very close formation back to our reef raft. Many years later Jack became the Assistant Secretary of the Air Force for Research and Development, and we worked together, never forgetting our shark encounter. In reality, the Johnson Island shark was probably nicer than many of the "sharks" in Washington D. C.

Our THV program began to have funding problems in the Pentagon and was slowing down. I had been tasked to prepare a report on another very highly classified program, a Ballistic Missile Interceptor, which made a lot more sense. The Army and the Air Force had been very quietly and secretly pursing this highly classified mission with support from the higher levels in the Department of Defense. The Army proposed a ground-based, RADAR-guided, high-acceleration missile to intercept the enemy ballistic warhead as it was reentering the atmosphere. The advantage was the atmosphere would separate the ballistic missile from any decoys accompanying the warhead to decoy interceptors away from the missile. The Air Force preferred intercepting the enemy missile in space, midcourse above the polar region, using a THV carried on top of one of our Minuteman ballistic missiles, similar to the Thor/THV concept that I was familiar with. This was the most effective solution, since any fallout would be above sparsely populated areas; however, it was susceptible to decoys. I was identified by someone who had access to both programs to provide a sensor/guidance report on a very sensitive issue on the Air Force approach. I completed the report and briefed the results to a group of scientists who enjoyed very high credibility with the very senior Department of Defense executives. The report was very well received and our discussions were stimulating. About a month after my briefing I was informed that the Army approach, which was less desirable but also much less risky, was selected. Our midcourse interceptor approach resurfaced with President Reagan 20 years later (Star Wars) after a great deal more research. My written Technical Report, which omitted all classified data, tied for first place for the annual Avionics Lab Kulpa award. This award and my stimulating briefing experience were soon forgotten. I was advised when I returned home that I had received an assignment to Vietnam to report to survival school within two weeks and then to Wichita AFB Kansas for F-105D training prior to being transferred to an F-105 Combat Wing in Thailand to fly 100 missions over North Vietnam.

I had, of course, been following the Vietnam War while I was absorbed in my space adventure; and, like many of my military friends, had been concerned about how we had arrived at this point. President Kennedy had successfully exploited the Army Green Berets and Rangers for several years against the North Vietnamese and Viet Cong in South and North Vietnam as well as Laos and Cambodia. These insertions involved small numbers of our best, highly trained fighters, which were usually successful. Indeed I would be very surprised if they were not successful— these were our pros, and the President and his Ivy league cabinet trusted them. Somehow or other they extrapolated that success into larger-scale operations in the South Vietnam jungles using Army forces against Viet Cong fighters who had 10 years' experience fighting the French. We had overwhelming military capability but not in the tropical jungles where our artillery, tanks and high-speed jet attack aircraft were useless. These assets would have overwhelmed the North Vietnamese military in the more European terrain of North Vietnam and a real war there would have been rapidly won. Apparently the President believed that if we engaged in North Vietnam the Chinese would come roaring across the border (ala North Korea-China) and a larger war would occur. Thus began the fatal strategy where we engaged the enemy Viet Cong guerillas in the impenetrable jungles of South Vietnam using draftees in ever-increasing quantities as our task grew more difficult. You cannot extrapolate success using small cadres of highly trained professional Green Berets to large-scale Army operations using draftees. When President Kennedy was assassinated President Johnson retained the same Secretary of Defense, Robert McNamara. After a naval episode in the gulf of Tonkin in mid 1964, McNamara devised his strategy of gradual escalation to North Vietnam. His strategy of gradual escalation involved starting with air strikes against low-value targets in North Vietnam, escalating to higher-value targets as dictated by their behavior. The principle air strikes were carried out by USN carrier aircraft flying off of aircraft carriers in the Gulf of Tonkin and USAF F-4D and F105D/F aircraft flying out of Thailand. By 1967 when I arrived, the North Vietnam defenses had been allowed to build up to the most formidable ever encountered. This was caused by the very restricted rules of engagement imposed by our politicians, coupled with the strategy of gradual escalation where our politicians selected terribly unimportant targets that would not unduly upset our enemy. We were forbidden to attack any targets in North Vietnam that were not approved by Secretary McNamara and they personally selected the F-105 targets in the Hanoi-Haiphong area. The result of this fatally flawed strategy was that less than sixty percent of the F105 pilots survived 100 missions. I was in deep yogurt.

Training for War Again: 1966

I had to leave home in December 1965, right after I had finished my report to attend our survival school near Fairchild AFB in Oregon. All aircrew who were operating primarily over enemy territory had to attend to learn how to survive as a POW, what to expect from interrogators, and escape and evasion techniques should we be able to escape. The course included many very stressful "torture" techniques allowed by the Geneva Convention. This was a refresher course for me.

The first week we had lectures, karate drills, and a great deal of physical exercise. Near the end of that week we were "captured" by enemy soldiers during a class with a lot of noise and shoving around. The next few days they worked us very hard with duck walks, lots of push-ups, very little food and lots of shouting. Sleep deprivation is always employed against POWs and we slept in a coffin that was leaning against a wall, so we always had our weight on our feet. The coffin tended to induce claustrophobia, so you had to concentrate on not losing your cool. Claustrophobia was their favorite technique and they used it effectively before your first interrogation. They forced you to squat down in a very small box and then forcibly closed the top, squeezing you into a squatting position inside the box with very painful pressure on your knees and calves. I was 76 inches tall, and I could feel great pain in my right knee when they closed the box because I believe they stood on the lid because I was so tall. Then they began to pound on the sides of the box to induce claustrophobia. I was in pain more than panic, and I concentrated on first orienting my body to a diagonal alignment and then rotating my body so that I was sitting diagonally in the box to relieve some of the pressure on my calves. Then I concentrated hard on anything other than the pounding. They let me out without breaking me into a panic reaction. My legs and right knee were very strained and my back was bleeding where I had forced my rotation to achieve a better position. They continued the excessive exercise, stressful interrogations, and sleeping in the upright coffin for several days until we were really sleep-deprived and every muscle ached. Then they put us all outside in an enclosure. It was December in the Oregon mountains and we had been issued our winter hat, gloves and jacket, so we knew this would be the start of our final exam— the escape and evasion for the last week. I was the senior officer, so I was the escape leader and the senior NCO was my enlisted deputy. We were all called to attention and the Camp Commandant came out and announced that they had been informed that an escape had been planned and that they were now going to punish the leader and his deputy.

They took the two of us out of formation and put us side by side in a two-man coffin, put the lid on, and placed us in a large grave. They shoveled rocks over the coffin but we had a vented coffin and rocks allow air. My deputy was a 25-year-old sergeant who was one of our professional combat air controllers and in splendid physical shape, but he was beginning to come unglued. He was hyperventilating and I could feel his body as stiff as a rod. I started talking to him getting him to relax his body and slow down his breathing and I finally cooled him down. It seemed like forever but they finally took us out of the coffin and grave and we were back in formation. In the middle of the night they freed us two at a time and gave us directions to our first check point with the requirement to arrive before sunrise. For the next week we proceeded at night, avoiding all trails and roads that were patrolled by our enemy. If we were caught, we would be interrogated until we broke. The incentive to evade patrols was very strong and we would walk about two-thirds of the way up the side of the mountain or hill over the natural terrain. It was mountainous and forested and we had more than a foot of snow. We would hide during the day and be very cold. At night we were working so hard walking through the rough wooded terrain that we would be soaking wet from sweat. I had a B-52 copilot as a partner and he was a prince. He was a 1st Lieutenant in good shape and kept up with me. We made it through the whole course went back to the barracks, showered, shaved, packed our duffel bags and picked up our graduation certificates. I did not feel grateful. I had a bad cold, had lost about 15 pounds, and my knee felt damaged.

There were six other pilots from my command, and they had all been on orders for the same F-105 tour that I was on. They all had found excuses to evade the orders and were en route to Vietnam to fly F-100s in South Vietnam where our USAF casualties were very low. When I returned home, I found that our command had received a quota for five F-105 pilots several months ago and had gone through more than 12 pilots who weaseled their way out of the assignment. Now I had it and had less than a week to start my training course. I had served overseas twice for a total of five years and only one of the 12 pilots who escaped the assignment had ever been overseas. There had been a lot of hanky panky in handling these five F-105 assignments. I wrote a letter to the Personnel Department reporting my findings and hoped they would correct their selection and assignment procedures. I was well qualified for the F-105, but three of the five assigned were transport pilots not qualified for high-performance jet attack aircraft.

Sabra and I discussed our situation. We decided that the family would stay in Dayton while I trained at McConnell AFB Kansas and then George AFB California. That would keep the children from transferring schools twice within the next six months. My most important decision was that my time was up on making a decision on religion. I was religious and had read a lot about philosophy and religion but had not attended church since my grandfather died. I had taken instruction to become a Catholic for years and felt guilty because my wife had a very strong faith and was raising the children as Catholic without my help and without complaint. I loved her very much and did not want to be separated after death, so I called the priest and asked my neighbor, Dr. Dan Knapke, to be my sponsor. Our families went down to the church and I became a Catholic. I felt very good about it and felt much better kneeling next to Sabra during mass. I have very rarely missed mass ever since. Our very excellent marriage became even better. She had inspired me with her gift of total faith and she was at peace with herself and her church. I prayed that I could become half that good.

I allotted all but $100 to Sabra every month so there would be no delay in pay if I were missing in action or a POW. I also refreshed my will and made certain that all of our assets were in joint ownership so that they would transfer to her without going through probate and the local county court system. Finally, I bought all of the declining term life insurance from the military associations that paid off for combat deaths. That was necessary to assure us that our children could attend college. I was shocked when I found out how little a widow would receive if I were killed on active duty. So I kept that insurance during the rest of my career.

The last thing I did was to go to our Foreign Technology Division where one of my AFIT classmates worked. He briefed me on all of the guidance data on the Soviet SA-2 Surface-to-Air missile and their analysis on each of our losses to the SA-2. I also obtained all of the employment and ballistic data on all Anti-Aircraft Artillery from 23mm up to 100mm. Most of our losses in North Vietnam were to AAA, then the SA-2 missile, and then the Mig 17,19 and 21. I copied all of the unclassified data and retained the rest of it in memory. Now I was ready to master the F-105. I left my family in Dayton and drove our 1964 Ford to McConnell AFB Kansas to report to duty as a Tactical Aircraft combat pilot.

F-105D Training: Early 1966

I was assigned to the 23rd Tactical Fighter Wing, 563rd Tactical Fighter Squadron, with 21 other pilots for qualification as an F-105D combat pilot. The 23rd TFW had been sent to Thailand right after the Gulf of Tonkin and had flown some of the very first combat missions in North Vietnam beginning in August 1964. Now the Wing was assigned the job of training replacement pilots for the F-105 combat Wings in Thailand. Most of our instructor pilots had flown 100 missions over North Vietnam. When the 23rd TFW returned, they were replaced by two permanent Wings, one at Tahkli Thailand AB, and the other Wing at Korat Thailand AB. It required four F-105 wings to support two wings in combat in a high loss rate war. We had two Wings in the States training replacement pilots, one at McConnell AFB Kansas, and one at Nellis AFB Nevada. Two Wings were required in the Far East to temporarily replace pilots and aircraft lost during combat before the next class of permanent replacement pilots arrived, one in Okinawa and one in Japan. The F-105 combat tour was one year or 100 combat missions over North Vietnam, whichever was shortest. This required six to nine months depending on the pace of battle and the number of bombing halts. The pace of battle for F-105s was controlled by President Johnson and Secretary McNamara.

We had several weeks of ground school. We studied and were tested on the F-105 aircraft and its subsystems in the morning and flew in the afternoon. I felt right at home on the flying and could fly tight formation and acrobatics immediately. The cockpit had more than 120 controls displays and switches. It could carry every kind of conventional bomb on the four wing stations, centerline station or the bomb bay and could carry a nuclear weapon in the bomb bay. We could carry sidewinder air-to-air missiles or Shrike RADAR Homing Missiles and/or ECM jamming pods on the wing outboard stations. We had a 20-millimeter Gatling gun capable of firing 6000 rounds per minute in the nose, an all weather RADAR, RADAR Warning systems, a Doppler navigation system, UHF Communications and TACAN/VOR/ILS radio navigational aids. Finally, we had an autopilot for en route navigation or autopilot ILS. The F-105 was one of the few aircraft that could fly at supersonic speed at sea level and we could carry a larger bomb load than the B-17 WWII bomber. It was the largest single-engine, single-cockpit fighter in the world, and in the next 16 weeks I would learn to master this beauty.

I had no problems flying the aircraft, and all of my tactical fighter skills came back rapidly. Two of our classmates had never flown anything but large transport aircraft and it was very clear that they would never make it through the course; and, if they did, they would be dead before they flew 10 combat missions. We were all relieved when they were eliminated and sent to another training base to fly small spotter planes in South Vietnam. Several pilots were from our Strategic Air Command and had been flying B-47s or B-52s and they took a little longer to catch on to the very responsive fighters and to being a one-man crew. They were used to a four- or five-man bomber crew, and we operated as a flight team of one, two, or four aircraft. The F-105 was equipped for aerial refueling, which was accomplished by flying your refueling probe into a basket connected to a hose from the tanker or flying your aircraft up to the tanker's boom to be stabbed by the tanker operator's probe. All KC 135 tankers were equipped with the refueling boom and the hose and basket. We all had to qualify on both the boom and basket during day and night conditions. On most of our combat missions we would have to refuel going to the target and returning home from the target. All of us qualified with only minor problems. By the end of the course all of us had qualified in dive bombing, strafing, and rockets, and had flown three nuclear weapon missions consisting of flying a low-level route under the hood (simulating an attack where the entire mission is flown in clouds) to a RADAR target and delivering a simulated nuclear weapon somewhere near the target. I had the best overall scores on dive bombing, gunnery and had the best scores on the nuclear attack. Only two of us had qualifying scores on the nuclear deliveries. The all-weather nuclear delivery in a single-seat fighter is very challenging and you are working hard the entire mission.

We flew four flights of four in a combat formation to George AFB California, with one aerial refueling to complete the air-to-air fighter tactics portion of the course. It was a lot of fun. I was well versed in this because of my F-86 experience in Korea. Once again, I had the best aerial gunnery scores and I had the privilege of leading the 16-ship formation back to McConnell AFB with one aerial refueling en route. This was our last F-105 flight in the United States. We had successfully completed our training, and now all 20 of us had orders to report to our combat Squadron in Thailand after attending a jungle survival school in the Philippine rain jungle.

We had a graduation party that night for our class, instructors, and the wives. The Squadron had lost several of their pilots during their combat tour in 1964-1965, and many of the pilots were missing in action (MIA) or prisoners of war (POW).

Their wives were waiting to find out their husbands' status. Our Squadron always invited all of these MIA and POW wives to all of the social affairs to keep their spirits up. It was very sobering to see the large number of these wives at the party and made you wonder how many of our wives would be left in this limbo. That was a defining moment for me, and I later became very active in our "Return our POW/MIA" effort. I packed my bags, checked out of the base, and drove home to my family in Dayton for a 30-day leave.

It was good to be home, and my family welcomed me with hugs and kisses. All of our neighbors expressed their support for us. The war had grown unpopular and the news was full of unpatriotic garbage that was rampant in Ohio and elsewhere in the U. S. There were many media people who were violently opposed to our bombing effort in North Vietnam, and I worried about people bugging my family. It made it very hard. Our good friends Ray and Mel Grazier and Chuck and Rose Lysaght assured us that they would be available to help Sabra while I was gone, and I knew that they would. The children were out of school so we planned to take our family to St. Louis to visit with my family and then to Pensacola to visit with Sabra's family. I would fly out of Pensacola to Camp Pendleton, California, where the military would fly me to the Philippines. Sabra stayed in Pensacola with her family until school started. She then drove our family home to Huber Heights Ohio. She and I had written each other every day when I was in Korea and we vowed to do it again. Her Aunt Wissie was a nun and had given me her Saint Christopher medal before I left for Korea, and I wore it on all of my flights. I would wear it every day, all day, on this combat tour. I kissed all of my family, loaded my baggage and myself into the cab and left for Pensacola Airport.

I arrived in the Philippines about three days later and joined my comrades on a bus taking us to the Jungle Survival School. Several of my F-105 classmates were in our group and after two days of school covering safety in the jungle, what was edible and drinkable, we were taken deep in the jungle and left to our own resources. It was easy compared to our trek through the Oregon mountains in winter. Major Rudy Fiedor, my F105 roommate in Kansas, and I teamed up and before the sun went down had each built ourselves a sturdy bed frame elevated above the jungle floor and protected from the rain by a palm leaf roof. Palm leaves also served as our mattress. We used sturdy, easily cut small trees as structures, notched the connecting points with machetes and tied them together with parachute strands. Water was everywhere and we drank rain drippings off of palm leaves. Food was harder to come by. We had sampled insects and raw snake and fish during the school but relied mostly on big seeds or unknown fruits that we found. We would tongue-taste everything before we ate it and discarded those that tasted sour or bitter per our school instruction. It was beautiful in the rain jungle and we slept well once we grew accustomed to the night sounds in the jungle. It was very humid. Generally, if you could parachute safely to the jungle floor and enemies were not near, survival did not seem to be a big problem. The top layer of the jungle was about 200 feet above the ground, the next layer about 150 feet and the bottom layer about 50 feet. Hopefully you and your parachute would break through the top and maybe the middle layers and become hung up in the bottom layer. We carried about 70 feet of lanyard and an elevator gimmick that you could lower yourself that far from your hung-up parachute and hopefully be able to reach the ground or climb down the tree from the end of your line.

We were being watched by natives and a military survival specialist and were debriefed after our adventure was over. Rudy and I agreed after it was over that it was fun and valuable. Valuable, because damaged F-105s were always trying to fly their aircraft back to the jungle before ejecting where the North Vietnamese enemy was rare and native tribes with CIA links would be rewarded for returning US aircrews. Also we had Rescue helicopters escorted by A1E fighter aircraft that had a wonderful record of picking up Navy and Air Force pilots from jungle areas in North Vietnam.

That was the end of my preparation for this combat tour. I had converted to Catholicism, arranged our finances for my family, learned how to be a POW, escape, evade and survive in the jungle, and best of all learned how to accurately dive bomb and kill targets on the first pass so that you did not have to go back the next day and try it again. I also had studied all of the enemy weapons that had the highest probability of shooting me down, and how they could do it. My greatest fear was being shot down and recognized by the Soviets assisting the North Vietnamese in North Vietnam as a valuable intelligence source. I would disappear into Russia and never survive the interrogations to obtain my special knowledge of Berlin electronics and our Satellite Interceptor Program. I would have to keep a very low profile and avoid having my identity revealed in any newspaper.

CHAPTER 13

Combat Tour, North Vietnam 1967-1968

I was assigned to the 355th Tactical Fighter Wing in the 354th Tactical Fighter Squadron at Tahkli Thai Air Base. As one of the senior Majors, I was the Executive Officer for The Squadron Commander and the B Flight Commander. Major Lonnie Ferguson was our operations officer and Major Jack Hart was our scheduling officer. I had known both of them and several other Squadron pilots who were at McConnell AFB Kansas in earlier Replacement Training classes. All of the senior officers quarters were full, so I moved into the Captains quarters.

Major Bob Kennedy
355 TFW 354TFS Tahkli AB Thailand
F105D (Thud) Pilot

My squadron commander took me up on my orientation flight, put me in trail formation and really put me through a wild bunch of maneuvers. I stayed with him easily and we returned to base. He had to know how well I flew the airplane, and I knew that he was pleased. He also had reviewed my bombing scores and Major Jack Hart also knew that my bomb scores were very good. This was a critical skill because the President of the United States picked our targets, and if they were not destroyed, we went back to that target again and all of the enemy gunners knew it and were ready.

President Johnson and his Secretary of Defense, Robert McNamara, devised the strategy of gradual escalation for the war in North Vietnam. President Johnson had stated the objectives of our bombing campaign in North Vietnam as follows: Impose a penalty on North Vietnam for violating the 1954 Geneva agreement to not interfere with the government of South Vietnam; reduce the flow of war supplies from North Vietnam to South Vietnam and; employ a large workforce in North Vietnam repairing bomb damage, thereby reducing the number of men available as war fighters. Nothing about winning.

Those very limited objectives obviously had produced very limited results. They had selected 100 targets of importance in North Vietnam and were attacking them in order of least important to most important. It provided a wonderful training environment for North Vietnamese and Russian missileers, gunners, and Mig 21 pilots in addition to their electronic intelligence people who could study our ECM tactics and strategies. Soviet arms and equipment flowed down through China to North Vietnam and thence by truck along the Ho Chi Min trails through Laos, Cambodia, and into South Vietnam. Our rules of engagement dictated by the President were that only the targets designated to be hit would be attacked and no others. Thus we could not attack the three airfields in North Vietnam where the Mig 21s were stationed that took off every day to shoot us down. Just one month before my arrival, our squadron strafed a Russian freighter in Haiphong harbor thinking that a deck gun had fired on them from the ship. We could fire on them if they fired on us. There were no guns on the freighter, so the gun was probably on the other side of the freighter. The entire flight of four were court-martialed. Two received letters of reprimand and our Wing Operations Officer was relieved from duty. Politics had entered the daily operations of the war fighter. The President is the Commander in Chief, so political guidance and objectives are respected but they now had invaded our operational strategy and tactics. I believed this interference in military operations was another sign of distrust of the military by President Kennedy's Ivy League cabinet, many of whom were now in President Johnson's cabinet. Further I believe the politicians' distrust of the military stemmed from General MacArthur's mistake in trying to force President Truman to extend the Korean War into China and use nuclear weapons to decimate the numerically superior Chinese infantry. President Truman denied the use of nuclear weapons. General MacArthur had already advanced his troops to the border, betting on approval. We were far from our supply line and the results were disastrous when the Chinese reacted by charging across the border, inflicting high casualties on our troops. I believe that was a key cause for the distrust of the military by our political leaders. Now, unfortunately, political restrictions are a fact of life and even our Congress now insert their restrictions into our battlefields. It was then, and is now, a recipe for failure.

Your first five missions were planned to be easy targets in the southern half of North Vietnam where there were no SA-2 missiles, Migs, or AAA guns that were effective above 5000 feet. It gave us experience in refueling, maneuvering and dive-bombing live munitions in a fully loaded F-105. It was required training because the fully loaded aircraft was less nimble and could be stalled if excessive maneuvers were made at low speeds. I learned that the fully loaded F-105 handled very well as long as you maintained a high energy level.

The very dangerous missions were in the two most northern areas around Hanoi, the capital, and Haiphong, the biggest seaport. The great majority of our losses occurred on missions to those areas. The average 100-mission tour only included about ten of those missions. They were scattered throughout the fair weather season but had been increasing. So had the casualties, and I now had my own room with the loss of one of our majors. We normally wait until a pilot has flown 10 missions before being allowed to fly in the two most dangerous areas. On my sixth mission I was scheduled for the first time with the main strike force. The strike force consisted of: four flights of four F-105D strike aircraft, two flights of F-105F Wild Weasel aircraft to suppress/kill the SA-2 missile RADARs, two flights of F-4E aircraft to protect us from Mig-21s, four EB-66 aircraft to jam enemy GCI RADARs, and about 8 KC-135s to provide refueling to the penetrating fighters going to and from the target. We did two of these major strike forces every day and I was on the early morning strike. Our primary target was a bridge near the Chinese border where we expected very heavy AAA but no Migs or SA-2 missiles. The alternate target was the Hanoi Railroad and Highway Bridge in Downtown Hanoi. Our Wing had just brought down the Paul Doumer Bridge in downtown Hanoi, and it was the first time that President Johnson had ever authorized a strike on an important target in the capital. That had been three days before, and our morale was high because we had finally been allowed to bomb a valuable target. The early morning briefing began at 3:30 a.m. and they announced that we were going to bomb the alternate target, the Hanoi Railroad and Highway Bridge. This was the other major bridge that carried all of the major highway and train traffic from China across the Red river. The two bridges should have been the first targets struck in August 1964, but they were not attacked for three years even though they best supported the President's stated objectives. I was flying the Strike Force Commander's wing and would be the second man in on the target. Downtown Hanoi on the sixth mission was going to be drastically different from my five previous missions.

We took off in total darkness and climbed to 15,000 feet while the other three flights of F-105 strike aircraft joined up into strike force formation; four flights of four F-105Ds in box formation. We spotted the flight of four KC-135 tankers ahead just as the sun peeked over the horizon at our altitude. These strike missions were planned so that the early morning strike flights would spot their tankers at exactly the first light of day and the afternoon strike flight would be able to finish refueling on the way home from their bombing raid as the sun went below the horizon. Thus most large strike forces refueled in daylight. This allowed two strike force missions against North Vietnam per day for each F-105D Wing; a total of four raids on North Vietnam with 16 F-105 bomb loads on four targets a day. A strike F-105 normally carried either six 750-pound bombs or two 3000-pound bombs. Occasionally, the first strike flight carried four cluster bombs to suppress the AAA flak sites in the target area. On major targets, both F-105 Wings would hit the same target separated by a few minutes. This morning, both wings were on the Hanoi Railroad and Highway target; and, as I remember it, we were first.

Each of the four flights joined up with their assigned tanker and we refueled. Each KC-135 had a two-ship F-105D formation at the end of each wing and we took turns sliding to the rear of the boom that projected out of the rear of the KC-135. When we were in position, the Boomer in the tail of the tanker would gently insert the nozzle into our fuel receptacle, which we had opened when we joined on the tanker. We would stay on the tanker's wing until we were deep into Laos. Then we would top off our tanks one more time and each flight would leave their tanker and rejoin into our combat box formation. The tankers would go into orbit on the Laotian side of the North Vietnam border and wait for our return from our mission. Their orbit was outside of the maximum range of the North Vietnam Mig-21. They were a very precious asset and as a flying fuel tank very vulnerable. We would increase our airspeed up to 480 knots and proceed into North Vietnam, arming up as we crossed the border, guns and bombs armed and checked, ECM pods turned on, and we would move into ECM formation. Additionally, I recommended my flight members to turn off their cockpit pressurization and switch their oxygen to breathe 100 percent at that point. Our cockpits were cooled and kept at a pressure corresponding to 8000 feet, using air from our jet engine compressor. We were usually flying lower than 16000 feet and when we turned our pressurization off, the cockpit would become uncomfortably warm like a greenhouse in the tropics. But if you were hit with your pressurization on, anything going through your compressor would come into your cockpit through your pressurization vents. Then you would be trying to assess and fix your battle damage with oil, hydraulic fluid or burning fuel all over you and your cockpit. Turning off your pressurization was uncomfortable, but it saved more than one life.

Our ECM formation, basically, was to take about a 500-foot vertical and horizontal separation between our four aircraft. Our ECM pods put out a signal that denied range information to their SA-2 missile guidance RADARs and filled their guidance beam with noise. Thus they would have to guide their missile into a blob of angle noise about 10 degrees in diameter. The SA-2 Missile crew knew there were four aircraft somewhere in that blob but they did not know where. There were four blobs of noise in close proximity, one for each of our four flights. All they could do was to launch one or two missiles into that blob and hope that one of the missiles would be pointing close enough to one of our aircraft to detonate the proximity fuse on the SA-2 missile. If they were lucky, the expanding fragmentation warhead might hit one of our aircraft. They also knew that when they launched their missile, all of our aircraft would spot the rocket plume with its trail of white exhaust and that our F-105F Wild Weasel aircraft would have detected its search beam, observe their tracking and guidance beam, and launched one of their Shrike RADAR homing missiles that would hit their guidance van shortly. When the Wild Weasel Crew spotted the tracking signal on their scope, they would broadcast "Valid Launch for the Force," and all eyeballs would scan the skies for the missile. They had a range of about 25 miles and the SA-2 missile was as large as a telephone pole. It would be traveling at Mach 1 or 2 by the time it reached the strike force. You could see the missile miles away; and, if it was a spot on your canopy that stayed in the same place, you were the target and had better move. If it had a line of sight rate on your canopy, it was going to miss you. If it was skidding toward you,it was being actively guided and you were in trouble. It was unnerving when the missile would go through your formations at mach 2. You just had to trust in your ECM and hope that the missile operator had turned off the guidance signal to avoid being killed by one of our Shrike missiles. His problems were not over yet, because that launch plume and exhaust trail revealed his location to 24 F-105D and F aircraft and he was fair game for a visual attack with bombs or guns. We worked hard to discourage the SA-2 Guidance crews, and many of us proudly wear the SAM KILLER patch on our party suits. Many of them, however, have THUD KILLER patches or equivalents. It was a deadly electronic game.

When we reached the mountains, we were now into Mig-21 territory and inside of their RADAR ground controlled Intercept network. We would be coming into their RADAR coverage, but our four EB-66 jamming aircraft would be hiding our aircraft for part of the time. We went into partial afterburner at the mountains and increased our airspeed to 540 knots. Generally, this is where the Mig-21s would attack. They would take off from Kep or Phuc Yen airfields and climb up to just below the cloud level paralleling our course coming toward us. As soon as their Ground Control Intercept RADARS could see us through the EB-66 jamming, they would vector their airborne Mig-21s to turn in behind us, accelerating up to max airspeed. They would pop up above the clouds, spot us, lock on and fire their missile homing on our hot tailpipe. Our F-4E flights were behind us and high, to cover our tails, and would pick the Mig-21s up on their RADAR before they made their turn, engaging them to thwart their attack on us. Occasionally one would get through and either shoot one of us down or cause an entire F-105D flight to jettison their bombs and do a hard break to avoid the missile. We shot down more Migs than we lost to Mig attacks, and the F-4Es shot down more than we did. That day, we did not encounter any Migs.

When we left the mountains, we generally were about five to fifteen minutes from our target. Now we were most likely to experience SA-2 launches up to the target area with very heavy 85mm and 100 mm flak barrages near the target. These high altitude airbursts were time-fused to explode at a certain altitude in a pattern of large explosions. Each round would explode with an orange flash spreading out into a black cloud filled with metal shell fragments. Usually they came in volleys from several guns covering a large area and aimed by a AAA fire control RADAR that predicted our position after tracking one or more of our flights. They were usually close enough to put shrapnel into several aircraft and occasionally destroyed one or more aircraft. Our highest losses came from 57mm and 37mm AAA batteries during our dive-bomb run and low-altitude flight away from the target. These were fused to explode on impact. After a certain time, they would detonate in a grey puff to prevent them from falling back into their territory. The sky would be filled with grey puffs that were 37mm and 57mm rounds that had already missed their target. The Soviets had invested heavily in defending the infrastructure, industrial and military facilities in North Vietnam. Our intelligence estimated there were more than 10,000 AAA guns 37mm and higher in the Hanoi and Haiphon areas. This day there were no SA-2 launches, and the high altitude barrage was not close to us. The F-105 wing from Korat was about five minutes behind us and the enemy knew that their target was the same bridge after we bombed it, so their AAA batteries did not have to guess where they were headed and when they would enter their dive. I believe they lost two F-105s that day.

Major Dale Leatham was our force commander that day and I was his wingman. We were assigned the 57mm AAA site on the bank of the river directly ahead of our dive-bomb heading. Thus they had the premium location to shoot us down because they did not have to change their azimuth angle; just aim above the flights and let them fly into their eight-gun stream of tracers. We rolled rapidly into our 45-degree dive and both independently put the target into the bottom of our circular retical projected onto our combining glass. The flak site was firing all eight guns sequentially. The guns were in a circle, so it looked like a pinwheel spouting tracer rounds at us. We tracked the pinwheel down until it was under our pipper, keeping the pipper on target until we passed 7000 feet; then we dropped our four 700-pound cluster bombs. Eight cluster bombs were on the way to the target. Each cluster bomb would open on a timer and hundreds of cluster bomblets would be spread out into about a 200-foot circle. If any one of the eight bombs were within 200 feet, all of the AAA crew and probably their unexpended munitions would be destroyed. If we dropped them before 7000 feet, the lethal radius would be larger; and, if we dropped them below 7000 feet, the lethal radius would be smaller. We had been in the middle of that stream of tracers from our flak site when we pulled out of our dive. We were lucky to have avoided a hit. Now we each started jinking furiously, changing our vector every four to five seconds, separated enough to avoid collisions with our random movements. We were below 5000 feet, well within the range of both the 57mm and the 37mm AAA. It took their projectiles three to six seconds to reach us, so we changed our vector every four to five seconds to avoid being where they had aimed. We were traveling well above 600 knots by now, headed for "Thud Ridge" and the protection of the mountains.

The worst thing you could hear during our high speed jinking escape was one of our wild weasels call, "Valid launch for the force," which meant a guided SA-2 was homing in on one of us. We would be spread out of ECM formation; the missile was probably coming up behind us, and we could not outrun an SA-2 missile. We were advised, if that occurred, to go into full afterburner and dive for the ground. We could exceed the speed of sound on the deck; and, as we accelerated, a condensation shroud would form around our aircraft like a funnel and proceed to close down into a small circle in front of our nose. We would hug the earth, trying to break the line of sight between the SA-2 and the guidance site or, better yet, hope that the missile would have to dive down to reach us, overshoot the pullout and hit the ground. I would experience that later, but not today. We had reached the mountains and rejoined our four flights in ECM formation and none of us were missing. Now we had to watch for Mig 17s or 19s, fly to the border, de-arm our guns, turn off our ECM, rendezvous with our KC-135 tanker who had been waiting for us, fly home to Thahkli, and debrief. I would pressurize my cockpit at the border and drink a lot of water through a tube connected to a big jug of ice water. I would always be soaking wet because of my depressurized cockpit. I had flown my first Hanoi mission across the Red River and was now a Red River Rat, an exclusive club for pilots who have flown a combat mission north of the Red River, and I did it on my sixth mission. As it turned out, it was the only easy Hanoi mission that I would fly.

During our debriefing we learned that when the Korat F-105 wing had attacked the bridge, one span was down in the river, so we had been successful. That was always a relief because if the bridge had survived, both wings would have to hit it again tomorrow. Revisits to targets were usually bloody. Korat had lost one aircraft over the target and there was no word on the pilot. Another Korat aircraft was badly damaged in the target area but the pilot made it into Laos, ejected and was picked up by the rescue helicopter. I could immediately see the advantage of being the first Wing on the target. The strike photos from one of the flights behind us revealed that our cluster bombs were tightly concentrated where the 57mm AAA site had been, so we had hit our target. I never liked to drop the cluster bombs because if you were off on your dive angle, airspeed or release altitude, they could open early and spread the small cluster bomblets all around the area into civilian areas of the city. I always erred on the side of a late opening, which gave a smaller but more concentrated lethal area. I always felt very bad about dropping the cluster bombs on targets inside densely populated areas because of the probability of civilian casualties.

Later, I noticed that our times of attack in the Hanoi area were really a function of sunrise and sunset. This basically meant that our early strike force bombed Hanoi about 8 a.m. local, with our afternoon strike about 3 p.m. The enemy knew when to expect us, roughly what our penetration altitude would be, and what our target would be at least half of the time. No wonder we had a high loss rate.

The next day we again had a railroad yard in the Hanoi area as the target. I was flying element lead in the fourth flight, which was the last flight to roll into the target. Everything was the same except for the target. This time there was a very effective 85mm and 100mm barrage right at our roll-in point. Our flight lead decided to zoom up over the flak barrage, which severely reduced our airspeed. We had to gently roll into our dive because our airspeed was so low. The other three flights had plowed through the flak and already dropped their bombs before we slowly rolled into our dive. Now all of the guns were on us and we had a very long dive with tracers all around us. I felt a strong yaw with a noise and knew I had been hit. I continued my dive, released my six 750-pound bombs and started jinking toward the mountains. All of my instruments looked normal. I told my wingman I had been hit and to check me out when we got to the mountains. He looked over and said he could see a gash through the back of my canopy and another gash along the top of my aircraft but no sign of a fire. A more thorough check at the mountain revealed that he could see a hole that went through the back of my canopy and exited my aircraft just aft of the fuselage fuel tank. The boomer on my tanker kept telling me about the hole and worried about filling the fuel tank to full. I took enough fuel to make it back to Tahkli and landed without any problem. A photographer met the aircraft and wanted to take a picture of me in the cockpit with a 57mm-size hole that had missed my head, entered the back of my canopy, and left a dark mark across the top of my main fuselage fuel tank, exiting the aircraft through the spine. It must have been a 57mm tracer to make the burn mark, and the fuse had failed to explode the projectile. If the fuse had worked, it would have given me a big headache and blown up my fuel tank. That was the first of my nine lives. Aunt Wissie's Saint Christopher medal was working. I would not pose for the picture or allow the episode to be published. I explained that my family was in Dayton in an anti-war town and I did not want them harassed. They honored my wishes. The truth was that I did not want the Soviets to know that I was flying daily combat missions over some of their detachments and make my connection to the Berlin electronics or Satellite Interceptor. I never revealed my past work to any of my Squadron mates or my concerns about becoming a POW. Now I had two missions across the Red River.

My next five missions were also in the Hanoi area and it became obvious that President Johnson was beginning to hit more valuable targets and to bomb the Hanoi and Haiphon areas every day. On one of the days after we left the mountains, we had more than 50 Sa-2s fired at us going through our formations. It must have exhausted their supply, because we went for several days with no missile firings. We also had several weird Mig encounters. We had some Migs chase us out beyond the mountains with several F-4Es chasing them, and an SA-2 shot down one of their own Migs. A Mig made a pass at one of our wild weasels and he jettisoned all of his wing tanks and broke hard into the Mig. The Mig exploded behind him, and we guessed he had run into the wild weasel's wing tank. We all prayed that the Mig pilot survived to tell his comrades about the new guided weapons that could be fired from the rear end of the F-105F.

We were beginning to hurt the enemy, but our losses had mounted and some of our attacks were poorly executed. On one mission we left the mountains but could not see the primary target because of clouds, and the Force commander decided to attack the secondary target when he was over the primary target and still could not see the target because of weather. He made a sharp turn toward the secondary target, which strung all of our flights way out behind him. I was in the fourth flight and we were way behind the third flight. We had rolled into a very shallow dive-bomb run too far from the target. Our Vice Wing Commander, Colonel Larry Picket, was leading our flight. All of a sudden an aircraft slid in front of me and behind Colonel Picket's aircraft. It was a Mig-17 or 19 and he was right in front of me; however, Colonel Picket was right ahead of the Mig, and if I fired I would hit the Mig and maybe Colonel Picket. Suddenly the Mig pulled up sharply. I tried to follow, but my full bomb load did not give me a chance. After we landed, my wingman, Captain Bob Bennet, claimed that he had shot down the Mig that was behind me, which could have been the wingman of the other Mig but nobody could confirm it. It all happened in a few seconds with no radio calls announcing the Mig attack, since there was so much chatter on our strike frequency. The Migs must have cut us off in the turn-away from the primary target. All four flights were strung out in a shallow dive toward the alternate target and our bombs missed the target. It was a sloppy mission and a miracle that nobody was lost. Unplanned surprises in the Hanoi area with four flights in ECM formation were a recipe for disaster.

We had a new Wing Commander and I knew him. Colonel John Giraudo was the operations officer in the 39th FIS in Korea. I was a Second Lieutenant and he left our Squadron and became Commander of the 25th FIS. He was shot down over North Korea and was a POW until they released our prisoners in late 1953. Several of us drove up to meet the returning POWs when they landed in Seoul, and he was greeted by a group of 39th and 25th FIS pilots when he arrived. He also was shot down in Germany during WWII and was a POW until the war ended. Since he had been a POW in both theatres, he could not be assigned to combat again, but he had volunteered and insisted he be assigned to an F-105 Wing, which had the highest loss rate in Vietnam. He was a seasoned combat leader.

He had a lot to catch up when he took over the wing, so it was a few weeks before he flew with us. When he started on the Hanoi missions, he flew as a wingman to get used to our operation. I was in the first strike force that he flew on a Hanoi mission. We had the usual short briefing and flew a mission that was marginally effective, followed by a very short debriefing. Colonel Giraudo had remained silent throughout and after the debriefing asked if anyone else had anything to say. Nobody responded and he stood up and told us that we were going to make significant changes in how we planned and executed our missions. He did not blame anyone but you knew he was unhappy and very shortly many changes occurred.

The changes were significant. The Force Commander was required to coordinate his attack plan with the EB-66s to be sure that we would attack coming out of their main jamming beam. The coordination with the F-105F wild weasels and F-4Es, which had always occurred, was strengthened. The KC -135 Tanker Pilots attended the Strike Force briefings and debriefings. The maintenance and armament chiefs became better coordinated and attended the briefings. Our mission planning improved and the teamwork between all participating and supporting organizations also improved. As a result, our combat losses decreased and the morale of our F-105 pilots improved because our confidence in the plan of attack increased. Colonel John Giraudo had made his leadership abilities apparent and we became a much stronger Wing.

We flew for 30 days and then took five days off. We needed the relief from the stress. I would catch our C-130 flight to Bangkok, and Lt. Colonel Sculley was stationed in the American embassy. Ray and Sheila Sculley and their teenage children were our friends during our tour in Wiesbaden, Germany. Ray would pick me up at the airport and take me to his plush apartment in the embassy compound. Sheila would cook a delicious dinner and I would thoroughly enjoy an evening with his family. The contrast was overwhelming but very welcome. The next morning I would leave, take the train to Patia Beach and rent a hotel room in the resort area. There were usually other F-105 pilots there on their five-day break. We would go to the beach on the Bay of Siam, rent a boat and go scuba diving or fishing for half a day. Happiness was lying on the beach and eating fresh bananas and pineapple that were picked at maximum sweetness. I have never again had a banana or pineapple that tasted that good. Young boys on the beach would run off and pick them and cut them up for a very small price. They were excellent. We would spend the evening in the bar. The hardest thing to do was take the train back to Bangkok, fly back to Tahkli, and check the schedule to see if I was flying the early morning or afternoon flight. I spent every five-day break like that, except when I shopped in the Bangkok jewelry shops for sapphire rings and gifts for my family.

The war tempo had been increasing. I had flown 40 missions, and 21 of them were in the Hanoi or Haiphong very high-threat areas. Before we arrived, very few strike pilots had 21 missions in those areas when they finished 100 missions. Our casualty rates reflected the increased number of targets our President was selecting in the high payoff industrial area. More than half of our class of 20 pilots who trained with me at McConnell AFB Kansas were already POWs, MIAs, or KIAs. We had kept both the Paul Doumer and Hanoi Railroad and Highway Bridges down in the Red River, struck the steel mill, several lesser bridges, and two airfields.

Every time we hit a valuable target, the North Vietnamese would return to the peace talks in Paris and we would have a bombing halt for a short time. When we renewed our bombing raids, they would have stocked up on their AAA ammo and SA-2 missiles and we would be flying very tough missions again. If we had four or five days of good targets, the flak and SA-2 launches would decrease in intensity, probably because their ammo and missile supplies were very low.

One of our troops brought back a French newspaper from Bangkok, and there was a picture of Jane Fonda surrounded by smiling North Vietnamese sitting in the gunners seat at a AAA site in downtown Hanoi. The woman is a traitor, and she and many others got away with it. During WWII she would have been tarred and feathered and run out of town. The media and the movie actors were on our side during WWII, but never since.

We worked about 14 hours per day, seven days a week with five days off every 30 days. Most of the time I flew the morning mission briefing at 4 a.m., returning from the mission about 10 a.m., going to the office to do my executive officer job and leaving the office about 6 p.m. I would eat dinner, write Sabra a letter, and go to bed. Major Ed Capelli and I would go to Catholic mass together every day that we could. I also would say the Lord's Prayer and Hail Mary prayer while we were refueling. My office job was mostly paperwork, but I spent a lot of time rewriting the officer fitness reports for our pilots who had completed their tour and were going home. Most of them were poorly written and omitted many accomplishments that would assist in their promotion. These were the troops who had shown up for the highest loss rate jobs in the USAF and they deserved to be promoted before those who avoided or evaded the assignment. I remember how upset I was when I read my reports for the first time in 12 years and found that many of my accomplishments were not included. We never had a full compliment of pilots and always had to borrow pilots from Japan and Okinawa to make up for our losses. Every one of our pilots had to be scheduled to fly every day. We had three full colonels that flew with our squadron and at least one of them had to be scheduled every day. Our scheduling officer, Major Jack Hart, had the most critical job in the squadron and was our "Rock of Gibraltar." He was also our most decorated combat pilot and led the toughest missions.

About half way through my tour, my Wing Commander called me in to his office and he read a Letter of Reprimand to me from a Major General in the Air Force System Command. The General was in charge of Personnel and he had taken offense concerning my letter alleging assigning unqualified pilots to fill his F-105 assignment quotas while allowing qualified pilots to be excused from the assignment. My Wing Commander was well aware that this was happening, because he was aware that some of his pilots were not well qualified to fly fighters. He read the letter to me as he was required to do and provided me with a copy with all of the endorsements. All of the comments from the Tactical side of the house were neutral and my Wing Commander noted that he had read the reprimand to me and discussed it. This was probably the kiss of death for my career, but I believe Colonel Giraudo thought that I had done the USAF a favor. He never mentioned it again. I was planning to retire in 20 years and only had three years to go, so it would probably only cost me another promotion. Oddly enough, I never saw any evidence of the letter again other than my own copy. If any of the endorsing officials read my letter and the attachments they would have noted that my attachment contained documents that confirmed my allegations.

Colonel Giraudo flew with all of the squadrons and usually at the last moment. Three times he took my flight lead and I flew his element lead. We would fly the assigned mission and go back to the tanker and take a full load of fuel. Then we would leave the strike force and go back into the edge of the Hanoi area, outside of the SA-2 range but inside of the Mig-21 area. He knew I was in the 39th FIS in Korea and could fly the ancient fluid four air-to-air combat formation. We would orbit the area trying to seduce the enemy Mig-21s to intercept us, but we were never intercepted. We would then go home to Tahkli, sometimes looking for random targets to strafe in the southern part of North Vietnam. The southern sections were much safer than the north unless you went below 5000 feet to strafe. Then the high rate of fire 23mm guns could be lethal. We lost two young Captains in the south on strafing runs. I always briefed my flights to never strafe unless it became necessary; however, if the Wing Commander strafed, I would follow him anywhere. We never logged it as two missions.

I have very vivid memories of only a few missions. Most missions in the southern part of North Vietnam or Laos were easy for the F-105Ds because we seldom descended below 5000 feet and very few places had AAA weapons greater than 23mm. The 23 mm was lethal up to about 3500 feet and very lethal if you chose to make low altitude strafing passes. I flew 48 missions in the southern portion of North Vietnam and four missions in Laos, which did not count toward my 100. The tough ones were the 42 missions in the Northern part of North Vietnam. Our group was the first group to fly so many missions in North Vietnam, because President Johnson was increasing the pressure on North Vietnam and I would guess that we flew about twice as many as prior F-105D troops. Major Jack Hart, who arrived a month before me, flew 55 missions up north, and his group were the only ones who flew more missions up north than we did.

My most memorable missions occurred in the September through December time frame. The worst one occurred en route to Hanoi about two minutes before roll-in. There was a high altitude flak barrage above our formation. Suddenly my aircraft pitched up sharply and I was high above my flight. I had put in corrective control and I suddenly pitched nose down into a dive. I was in a porpoise maneuver and could not control my aircraft. My stability augmentation system was obviously not working, and I had hit the Stab-Aug on switch several times without results. I checked all of my circuit breakers and none had popped. The aircraft continued its crazy maneuver and my oscillations were increasing. I was going to crash unless I regained control. I dropped my dive breaks and reduced my power, which seemed to help, and the oscillations stopped below 250 knots. The Strike Force had rolled in, bombed the target, and were making a high-speed exit. My wingman, Major Iverson, stayed with me, keeping his speed up but orbiting above me. I regained control about 4000 feet above the Red River in downtown Hanoi and I headed for Thud Ridge. Even flying at 250 knots, the aircraft flew like it was balancing on the tip of a needle; very unstable. I kept trying to increase my airspeed, because every gun in Hanoi was shooting at my wingman and me. The best I could do was to climb at 250 knots with 37mm and 57mm tracers crossing close in front of me. The gunners had never seen an F-105 flying at 250 knots, and all of the rounds were crossing in front of me. This was my worst nightmare—ejecting into Hanoi and becoming a POW. It seemed like forever to start my climb up Thud Ridge. I was a sitting duck for Migs or SA-2s but none were fired and Ivy stayed with me. We climbed up to our egress altitude and I could still control my aircraft. Ivy checked me and could not see any battle damage. I tried everything I knew to re-engage my stability augmentation system to no avail. I had dropped my bombs above the mountains, hoping that this would not trigger the oscillations. Our tanker had waited for us and I knew the rendezvous and refueling might be risky. My tanker pilot started his turn into me several miles out; and, when he rolled out, he was directly ahead of me, about a half-mile—a magnificent maneuver, which reduced my join-up problem. He slowed down and I took on enough fuel to get to Tahkli. When we arrived back at the field, I carefully performed my pre-landing checklist at 15,000 feet and could maintain control at approach airspeed down to round-out airspeed and that was good enough. So I entered a flame-out pattern at 15,000 feet and landed safely. I was never so relieved in my life but I had lost a little bit of confidence in my F-105D.

It was hard to fly my next mission but within a week my comfort level returned.

On my 40th mission we were assigned to the first strike against the main airfield in downtown Hanoi—Phuc Yen. Their air defense command operated underground beneath the concrete runways and parking surfaces, and we carried 2000-pound bombs for the first time. We were after their Air Defense Command facility but would also be destroying their runways. The 2000-pound bombs could be fused with a short delay and were structurally able to penetrate the concrete and then explode. It was a much better weapon than the 3000-pound bomb, which had a weaker case and had to use a proximity fuse because the case would break up on contact with the ground and a muffled explosion would result. We should have carried 2000-pounders for most of our missions but apparently there was a shortage of these excellent bombs and plenty of 3000-pounders. Major Ed Capelli was the strike force commander and I was his wingman. We had just passed the mountains and were about two minutes from roll-in when the weasels called, "Valid launch for the force." I scanned the horizon looking for the rocket plume followed by the white exhaust trail and saw nothing. An Sa-2 detonated between us, followed by another that detonated directly below Ed's aircraft. I can still see it in slow motion; the fragments from the warhead expanding into a cone below his aircraft, with many fragments emerging from the top of his aircraft trailing fuel streams that erupted into flame. I called for him to jettison his bombs and turn back toward the mountains, and for the element to fly high cover for us while I escorted him to a safe bail-out area. Major Jack Hart assumed the Command of the strike force and they proceeded into the target. Ed's cockpit was black with smoke, so he had not turned off his pressurization. I called him to turn it off but about then he jettisoned his canopy. I thought he might eject over the Hanoi area but he did not, so he must have jettisoned his canopy to get rid of the smoke in the cockpit. His aircraft was trailing a huge amount of flame from his ruptured fuel tanks, but his engine was still running. We were losing altitude and I told him to climb for altitude and he started a climb. He could hear us, but we never heard any transmissions from him. We were about halfway to the closest safe bail-out area when the element leader started yelling. "Get Out!" Ed ejected. I could not see why he should eject unless his engine quit. My plan was to go until the engine quit and then zoom up to gain altitude and a little more distance and then eject. As it was, he ejected before he arrived in a safe area. In another five minutes we would have been in Laos where rescue was likely. I called for a rescue effort to pick him up, but they could not survive in that area. I went back to where he ejected and saw his chute before he hit the ground. I tried to contact him on his portable survival radio but there was no answer. I was orbiting his location looking for him, and for enemy troops, and making repeated calls to him. I thought I saw his parachute in a depressed area and went down to look closer. I knew I had been hit again by enemy fire but no warning lights came on in my cockpit. I climbed back up to 5000 feet and tried again to contact Ed. This time I heard a voice speaking in another language and I figured he had been captured and they had his survival radio. I rejoined the rest of my flight in high orbit and we headed for our tanker. I was very low on fuel and advised my tanker. He must have headed directly toward us, well into very dangerous territory for that flying gas tank, because when I hooked up I was on fumes. That Tanker crew had saved me from a bailout. We all admired our tanker crews, who registered many saves like this. We returned to base much later than the rest of the force, and I was called into the Wing Commander's office. Hanoi Hanna, the North Vietnamese propaganda newscaster, had broadcast that the F-105s had bombed Hanoi today and killed an Indian Guard at the Indian Embassy. Ed and I had jettisoned our bombs right after we were hit, and they figured that it must have been our bombs. The Generals in Saigon were upset and anxious for details. Fortunately, I had carried my own voice recorder to have my own record of my missions for debriefings. We played it back and could correlate where we were when we were hit and where we jettisoned our bombs. There is no way they could have glided that far. I had lost my Catholic mass partner and my friend. It was a bad day. Ed Capelli was listed as MIA for many years until they found his body more than ten years later.

Not long after that mission, we were once again targeted against the Paul Doumer Bridge, since they had patched it up sufficiently to resume traffic over that very critical transportation link. So we were going back to downtown Hanoi again. Major Jack Hart was the strike force commander and I was his wingman. We passed the mountains without encountering Migs; and, as we were progressing toward the target, an SA-2 was launched from at least 15 miles away. The weasels must have launched a Shrike at the site because the missile was not guided and failed to come close. However he had left a beautiful white plume from his launch location right near the foot of our bridge target. Jack maneuvered the strike force so that they had a good steep dive on the bridge. He and I were carrying cluster bombs for flak suppression, but both of us dove in on the well marked SA-2 site. We were in a near vertical dive when two SA-2s were launched and were coming straight up toward us. It seemed like a long time that we were rapidly charging toward each other when suddenly both missiles went haywire off to the side. I remembered that the SA-2 could not be launched straight up, so he must have launched at his maximum launch angle and then tried to guide the missiles to a steeper trajectory and they had stalled out. All eight of our cluster bombs covered his launch site and Jack and I earned another SAM killer patch.

My last strike mission on the Paul Doumer Bridge was 18 December 1967, and I was in the last flight on target. Our flight leader rolled in on the wrong bridge! The three remaining flight members proceeded up the river at about 9000 feet, approaching the correct bridge at a 90-degree angle. We were spread out and in a shallow dive and could see that the bridge was still standing and the other three flights had not brought it down. All of us dropped our bombs from this totally non-optimum angle. The post-strike photo revealed that one of us dropped the bridge. Miracles happen.

I finally reached my 95 missions in early March 1968. The custom was to give you five easy missions for your last five, because everybody became very nervous when they were approaching 100 missions, the go-home point. They were nervous because not many of us were arriving at that magic number. Out of the 20 of us that were in my class at McConnell AFB, only five came home with 100 missions; two came home six and a half years later after spending that time as POWs in Hanoi; and the other two Majors came home with medical disabilities. Our casualties were higher because we had a much greater percentage of missions in the most dangerous areas of North Vietnam. My last mission was an unimportant target near Diem Bien Phu. After we had dropped our bombs, we were diverted to an emergency action near the North Vietnam border with Laos. North Vietnamese troops were scaling the mountain where one of our covert electronic sites was located and we were under attack and outnumbered. We only had our 20 mm Gatling guns, and we rendezvoused with an unarmed Forward Air Controller (FAC) in a light plane. We could see the attackers climbing up the steep sides of the mountain and the FAC told us that our friendlies were at the top. We made individual passes from the back

My B Flight. All of us made 100 missions! Front: Capt. Lunde, Maj. Winters, Capt. Wright, Capt. Sisney Back: Maj. Iverson, Me, Capt. Scott, Capt. Cleveland

of the attackers, and our FAC reported good results. But after two passes we were out of ammo and went home. When we left, there was a flight of AlEs en route. I hoped they arrived on time. It was 14 March 1968, and I left the next day for home. The President declared a bombing halt the next day, which lasted for a long time, so once again I had lucked out getting my 100th on the last day before the bombing halt. I had been recommended for the Air Force Cross, five Distinguished Flying Cross and twelve Air Medals but only one DFC had been presented when I left Thailand. I was very proud of the fact that in spite of the heavy losses in our Squadron and Wing, my flight had not lost anyone during my command. I did not realize that until one of my troops told me before I left. Apparently it had become almost a mystical anomaly within the Squadron and nobody wanted to discuss it for superstitious fear of breaking the charm. In our high-loss-rate environment superstition was rampant. I do not consider my St. Christopher medal a superstition, but I had one flight suit that I would not wear because I was hit the first three times I wore it. My St Christopher medal must have been all-powerful, because it brought my entire flight home.

I was very disappointed in our President's limited objectives and timid strategy of gradual escalation. We should have fought the war in North Vietnam where our superior firepower and technology could be exploited and not in the jungles where our draftees were fighting a guerilla war. The soviet supplies coming down through China could have been stopped at the Red River by taking down two bridges, three airfields and the Port at Haiphong. Their Air defenses would never have been allowed to build up to the excellent system that they built while we bombed unimportant targets selected by our President. We waited three years before we began hitting targets that hurt. The very limited war lasted more than ten years. We killed more than one million guerilla and North Vietnam enemies, the South Vietnamese lost more than 400,000 people and we lost more than 58,000 military. During these ten years, the United States had lost more than 120,000 people to drunk drivers, 200,000 people to murders, and a countless number of fetuses to abortion. Seven years after I came home our Congress, urged on by our anti-war groupies, withdrew the funds that supported the war, and our tarnished President Nixon had no choice but to pull out of Vietnam, deserting our South Vietnamese allies. It was our first show of weakness that would encourage future enemies to take us on. North Vietnam, a poor country the size of one of our states, had defeated the most capable country in the world with a constant flow of supplies from the Soviets, the assistance of our limited political objectives, and a weak congress.

Our very limited objectives had worn out the American people. Power is the product of national capability times national will, and our national will was zero. It revealed new strategies for our enemies.

It was a lousy war, but I am very proud to have served with the most dedicated USAF and USN pilots who repeatedly flew into North Vietnam against the most formidable air defenses, trying to execute a fatally flawed strategy conceived by our timid politicians. We obeyed their rules of engagement, suffering high losses as a result and ultimately retreating in defeat. I thank God that I survived, and I continue to pray for all of those brave aircrews that we lost.

My 100th Mission
Colonel Larry Picket congratulates me

CHAPTER 14

My New Mission

I left Tahkli on the morning flight to Bankok and caught a military transport aircraft to California. I took a cab to the San Francisco Airport and bought a one-way ticket on a red-eye flight to Dayton Ohio. Upon arrival in Dayton the next morning I caught a cab to my home. Sabra knew I was on my way but did not expect me on that day. I rang the doorbell, and when she opened it we both embraced and tears flowed. I had been on edge for 16 months and now I knew I was home for good and it was a wonderful feeling. Our kids were playing in the yard out back and we went around back and called them. They came running across the lawn and we scooped up all three of them and enjoyed a group hug. We were all delirious with joy, and our collie finally had his shepherd back.

My homecoming was really heartwarming. All of our neighbors and their children had watched out for my family, sheltering them from anti-war nuts. The newspaper headlines occasionally revealed how many F-105s had been lost the day before, and the kids in school were very conscious about combat losses in North Vietnam. My children were in the local Catholic grade school and before I went over, one of our parish members, Captain Brendan Foley, was declared MIA in North Vietnam and his six children were friends of my children. He was an RF-101 reconnaissance pilot and they flew over and photographed the intended target in North Vietnam before and after the strike force attack to reveal the damage. This of course revealed our line of attack before we attacked, which gave them a strong hint on our target and arrival time but showed the President evidence of the actual damage we had inflicted. The RF-101 loss rate was comparable to ours. The grade school all prayed for me because they all had felt the loss of Captain Foley. Ray and Mel Grazier and Chuck and Rose Lysaght helped Sabra and the children frequently but I could tell the strain that my family had been under for 16 months. They never found Captain Foley. Shortly after I returned, Ray was sent to Vietnam to fly F-4Es in the famous 555th Tactical Fighter Wing (Triple Nickel). We took care of Mel and her children while he was gone, just as they had cared for my family. He survived his tour.

I had to control my negative feelings toward our politicians and our media to continue my military career. I intended to retire at 20 years, just two years away. I had left Thailand on 30-day leave, awaiting orders for my next assignment. I was very concerned with our combat losses in North Vietnam because a conventional war against the Warsaw Pact in Europe would be worse against their more sophisticated defenses; and, of course, the Soviets had closely observed our tactics in North Vietnam, so our tactics were already compromised. Our next generation aircraft must do better with lower losses. My space defense experience had been pure physics and logic. I had a target set (satellites and ballistic missiles) and an error volume of uncertainty. My job was to find an appropriate sensor, resolve the tradeoffs between sensor, guidance, fuse and warhead to optimize the probability of kill, and then design the vehicle to contain these vital variables and interface with the launch environment. Now my desired mission was to do the same for USAF aircraft. I decided that I wanted to go back into the Air Force Avionics Laboratory and work aircraft avionics. I called my old boss, Jerry Pasek, and he said he would try to arrange it. This was not the brightest idea in my life because I was going back into the command where the two-star General, the head of Personnel, had given me a letter of reprimand.

Air Force Avionics Laboratory: 1968-1970

I received an offer to be assigned to the Pacific Command in a staff job in the group that passed the targets down to the F-105 Wings as an exceptionally qualified officer. This surprised me because it was a formal recognition that I was "special." I turned it down, because I wanted to pursue a higher purpose than passing the President's targets down to the troops who had to execute them. Meanwhile, the Avionics Laboratory commander had found an ideal slot for me—program manager for advanced development of aircraft navigation (666A), missile guidance (679A), and sensors (698DF). This was a job where I could help shape the avionics and weapons for the next generation of USAF aircraft. Jerry Pasek would be my technical director. I was very pleased because it was an ideal last assignment for me in the USAF.

I was notified that my combat awards had been received and would be presented to me in our big conference room with all employees present. They presented me with a silver star, four distinguished flying crosses, and twelve air medals. The pentagon had downgraded my Air Force Cross to a Silver Star and one DFC to an Air Medal. I couldn't care less. I was alive and well and that was reward enough. Then, because of the publicity, I was detached to the Treasury Department to go on a tour in Ohio to encourage people to buy Treasury bonds. I could not get out of it.

Award Ceremony, Wright Patterson AFB Ohio
Commander AF Avionics Lab presents Silver Star,
Four Distinguished Flying Cross, and Twelve Air Medals

I departed Dayton with a Treasury Department executive in full dress blues headed to Cleveland for our first TV interview scheduled for late afternoon. We arrived at the station and were escorted into a studio with a young good-looking news anchor. He explained we would have an informal conversation to get me used to the two TV cameras and then go into my treasury pitch when we were comfortable. He was made-up and very articulate. The cameras were on and he started our conversation. "Major Kennedy, you just completed 100 missions over North Vietnam. The Paris Peace talks reconvened today and North Vietnam claimed that they have won the war of world opinion because of the atrocities committed by the F-105 pilots bombing North Vietnam. Tell us what they meant by atrocities." I was furious but told him I was there to talk about our Treasury Bond drive. The obnoxious arrogant shrimp continued to press on his propaganda agenda, and I foolishly engaged in a spirited manner for about 20 minutes, and then we left the studio. I did not shake hands with any of them when we left. I angrily told the Treasury Representative that I was set up and very angry. We then proceeded to a reception in my honor with some of Cleveland's finest. I enjoyed meeting Bob Fellows, the famous Cleveland Indians pitcher and several other local VIPs. We went back to the hotel early to watch the 11 p.m. news. They presented about a three- or four-minute clip of my interview. All of my Treasury Bond pitch was missing, and they had cut and pasted the camera video from the two cameras, with my voice tapes slanted in such a manner as to imply that our bombing campaign was undisciplined and callous. I was furious and wanted to cancel the rest of the tour because this weakling from the treasury was incapable of screening the anti-war crowd out of my schedule. I had just finished my war overseas and now had to fight our enemies in the U.S. I refused any more press interviews and made my appearances in factories and businesses. I made the mistake of showing up at a local college where chicken blood was thrown at me. (It missed.) By this time I was disgusted with the anti-war crowd and their behavior. The factory workers and business people, however, were a joy to talk with. The fringe groups were rude, loud, and wrong. I never did it again.

I assumed the informal role of local leader of the Red River Rats Fighter Pilot Association at Wright Patterson AFB Ohio. It consisted of any pilot who had flown one mission north of the Red River in North Vietnam, and we had one mission—to work for the return of our comrades who were POWs in North Vietnam. We had annual meetings at various fighter bases in the U. S. and sought to bring political pressure on our politicians to achieve our objectives. We personally financed our operations and printed up post cards to send to the North Vietnam delegates at the Paris Peace talks to enclose in all of our Christmas cards. We also sought publicity to motivate our politicians to become more active on returning our POWs. We probably exceeded our political boundaries, and soon we received political pressure from Congress down through the chain of command. Although we were an informal ad hoc group, we understood the message: Knock it off. One of our brighter fighter pilots had a good idea—pass the baton to the wives of our POWs/MIAs. They took off with the POW bracelets and attracted a lot of publicity but no results from our politicians. Then a true American hero, Ross Perot, picked up the cudgel and the rest is history. He chartered a commercial aircraft and flew Christmas presents and several POW wives to Hanoi to deliver to our POWs. Naturally the North Vietnamese refused him and soon became very sorry that they had allowed him to come in the first place. It was a propaganda victory, and he personally financed the POW/MIA wives to expand our support from the American people. I am certain that his intervention resulted in the eventual return of our POWs. When they returned, our POWS related that their treatment improved after his visit to Hanoi, and our government finally began to pay attention. He will remain a true patriot and a real hero to all POWs and Red River Rats until the end of our days.

I had thought there must be a better way to do an air war all during my combat tour. First we assigned 16 F-105D dive-bombers to attack one target. If it was an important target, both combat wings were used for a total of 32 F-105D dive-bombers. We needed two RF-101 reconnaissance aircraft to photograph the target before and after our strike, four or eight F-105F wild weasels to suppress or kill the SA-2s, eight F-4es to protect us from the Migs, and four EB-66s to jam the enemy RADARs. Finally, we needed eight KC-135 tankers to refuel all of us going in and coming from the target. That is a total of 46 to 62 aircraft and crews to strike one target. Each F-105 Wing in combat could do two major targets per day. We had two F-105 wings in Thailand in day-to-day combat. We needed a total of two F-105 wings in the states to train replacements pilots and two more Wings in Okinawa and Japan to provide replacement pilots and aircraft when our losses were high. That is a very high number of committed resources and a very large number of aircraft sorties to destroy two to four targets per day.

If each attack aircraft carried four guided munitions, we could safely assume three targets per sortie instead of 16 or 32 aircraft sorties per target. That equates to 48 targets destroyed for 16 aircraft sorties with high confidence and much less collateral damage. The pentagon and Congress were currently averse to the high costs and low reliability of aircraft avionics and guided weapons, so we had not developed any. Our space people had already made great gains in avionic reliability and missile guidance and clearly the cost-effectiveness of three targets per aircraft sortie overwhelmed our current approach. Further, we could launch our guided weapons at a greater distance from the target, reducing loss rates and, best of all, with accurate guidance use smaller warheads and suffer much less collateral damage. The problem was that we did not have any non-nuclear guided missiles or bombs for ground attack. I thought it was a slam-dunk and made presentations to various military and industry groups meeting with some modest enthusiasm, mostly from industry and the scientific community. I began focusing our advanced development programs toward that goal, and we picked up more support from other research and development programs along the way. Our command had a Mission area planning group, and they had just completed an analysis of our future needs in tactical aircraft. It was excellent and provided a baseline for us to pursue the technologies that their studies and analysis revealed as potentially high payoff.

We started at the target and worked backward to the pilot in the cockpit. First we had to double the availability of our attack aircraft from day only in clear weather to day/night and hopefully all-weather. The 698DF Advanced Development Program already had demonstration projects on RADAR, Laser, LWIR, MWIR, and Low Light Level TV to determine their relative effectiveness as tactical target sensors, and we soon had more funds to accelerate the progress. I emphasized two programs on improving all-weather bombing accuracies using ground beacons and the aircrafts inertial platform with the digital data from both sensors integrated in a Kalman filter to achieve greater accuracy. My project engineer was 1/Lt. Bill Crabtree, who had recently received his doctorate in Electrical Engineering from Stanford. He was our first engineer with digital electronic skills. I had started the digital RADAR program on my prior tour in the laboratory and now had a digital electronics expert. He is brilliant and we continued to stay in contact for the next 36 years.

Our team accomplished a lot in two years. We built the first all digital RADAR test bed, using the RADAR that we had used in the B-26 to demonstrate the look-down mode that went into the F-15. We made it into an all-digital system, including the displays and installed it into an old B-47 using an F-111 radome on the nose. We then demonstrated two electronic phased array antennae. This allowed us to form and shape a wide variety of RADAR beams limited only by our transmitter power. We were no longer constrained by mechanically scanned antennae and could program and process interlacing electronically scanned beams. Thus we could interlace the air-to-air search, the ground map, and the terrain following RADAR modes simultaneously. This became the baseline design concept for the B-1B and B-2 RADAR with solid-state arrays instead of phase shifters in the early 1980s. With further improvements in the 1990s, it became the RADAR for the F-22 and F-35 fighters.

Our Long Wavelength Infrared imaging sensor technology became the primary night sensor for finding enemy targets for C-130 Gunships, Army Helicopters and, much later, USAF fighters. We demonstrated the use of several sensors and processing the data through a Kalman filter for the first time, courtesy of 1st/ Lt. Bill C. Crabtree. Our high-precision electrostatic Inertial platform proved to be the most accurate aircraft navigation system in the world and was installed in the B-52G/H and initially in the first F-17s. We mixed the inertial platform data with Doppler and LORAN data to obtain remarkable improvements in aircraft navigation sufficient to use for bombing. Later, this Kalman filter technique using two complimentary sensors is being used today (Global Positioning System and Inertial) to guide most of our precision-guided missiles and glide bombs. We demonstrated correlation guidance and radiometric guidance, which were later used for all weather missile guidance against ground targets. It dawned on Bill Crabtree and me that digital electronics had arrived and was the future, but a lot of architecture work had to be done first.

Our Lab was organized by application. Navigation, Electronic Warfare, Reconnaissance, and each division were working on an integrated (application) system without interfaces and ignoring how we would transfer this plethora of new information to the aircrew in the unlikely event that these highly integrated, software-dependant, complex kluges would play together. I was certain that we could greatly reduce cost and weight by time-sharing common apertures, displays, computers and processors between all of these applications. Further we needed a common set of interfaces, protocols and standards to function together. I recalled my F-105 cockpit with separate displays and controls for each function with the result that I had counted 125 different displays, controls and switches. The new threat environment with Migs, SAMs, ECM, and Flak were very distracting and made it hard to focus on the target. All of our divisions had done a good job bringing even more valuable information from the target area into the cockpit. Now we had to use this data, correlate it into the aircraft reference frame, perform computations and present the pilot with appropriate options at the right time to facilitate target kill while maintaining threat awareness. Finally, we had to design the cockpit controls, displays, and switches to match his physical abilities, sensors (eyes, ears, feel) and his brain. I formed an informal working group with scientist and engineers from the Flight Dynamics Lab, the Materials Lab, and a neurologist from the Medical Research Lab to see if any of them were working in areas relevant to this problem. They were, and we decided to better focus some resources to learn more. For example, our displays were black and white. The eye can only discern 10 shades of gray, which would never be sufficient to display our multiple categories of information. Multi-color displays gave us a much greater range of information. The Flight Dynamics Lab was already working this problem with the Materials Lab. The neurologist and I went way out on the human brain and senses. It turned out that we knew very little about the brain as far as the thinking processes, but we knew that it was a great deal more capable and probably underutilized. We both were amazed at how rapidly children increased their computer game skills with repetition and training. It gave us hope. We strayed way off the subject frequently—both of us were interested in mental telepathy, dreams and other unrelated subjects. He was a blithe spirit. Our common goal was to design an Avionics system where all applications would share common supporting navigational data, displays, apertures, processors and power supplies, and be structured with an architecture that facilitated new technology insertion. We also recognized the need to improve the reliability of our avionics, and cooling appeared to be the most promising approach. We eventually called our informal kluge the "Digital Avionics Information System" (DAIS).

Just as things were becoming interesting, I was offered the opportunity to go to the Air War College. I had just been promoted below the zone to Lt/Colonel, which surprised me because of my weird career. I figured it was because of my combat tour. I wanted to stay in the Lab because I loved my job and could contribute more in this position. Therefore, I turned down the Air War College. The personnel officer called my Commanding Officer, Colonel Kulpa, and he called me in to talk. He explained that this was an excellent opportunity since I was offered the AWC as a brand new Lt/Colonel and usually you had to be a Colonel to attend. He said that it meant that I was on the track to make General or be assigned to a General officer's position. I was flabbergasted. Did somebody tear up my letter of reprimand? Sabra and I had planned for a 20-year career with a second career, and I had found my niche as a scientist/engineer. I went home, and Sabra and I had one of our serious talks.

We had saved our money and put aside funds every month so that our children could go to college. Our plan was to retire after twenty years, and we would receive half of my base pay as retirement income; however, I knew several aerospace companies who would hire me in their Research and Development Lab. I had never discussed this with them because public law has rigorous rules about relations with your contractors, but I knew it would be a much higher salary than I would make as a Lt/ Colonel. If I went to War College, we assumed I could stay in for another eight years as a Lt/Colonel, as worst case, and still have enough to send the children to college if we retired in eight years and I could find a well paying job. The risk was that in my current job I was on the leading edge of military digital avionics, probably the dominant leader within the USAF. After War College and higher-level assignments, the rapidly moving technology would have made me obsolete. We were having a tenfold increase in our digital performance every five years. As a technology planner, I had to make very optimistic assumptions and we were exceeding them. It was exciting, but we loved the USAF life and our friends. We decided we would go to the Air War College in Montgomery, Alabama. I continued to stay in touch with my informal group in digital avionics for the rest of my military career. As I write this book 37 years later, all commercial and military aircraft are exploiting the fourth generation of our modest proposed architecture.

Air War College: 1970-1971

We rented our home in Huber Heights to a USAF Colonel and moved to a rental home in Montgomery, Alabama. We succeeded in registering our children in the Catholic school and I reported for duty as an AWC student. We were overjoyed to be with many of our comrades from other past USAF tours. Several of my Vietnam combat buddies, some of our European friends, and even a few pilots from my Korean tour were classmates, as well as recent friends from Wright Patterson AFB. The wife of my close friend from my recent combat tour, Major Ed Capelli, was there. Ed had been declared KIA and she had met and married another POW from our squadron, Colonel Dewey Waddell, who had been shot down and captured shortly before Ed and I had arrived in the squadron. She had known me only because I wrote the letters to the wives/families of our squadron mates who had been shot down, relating all of the known facts to them. She was very active in the POW/MIA wives' activities and was instrumental in obtaining a US Postage stamp for the POW/MIAs before their release. Several of my new classmates were also Red River Rats and we continued, along with the local POW/MIA wives, to focus attention on the plight of the POWS and MIAs.

The Air War College was the final professional school for high level DOD and CIA civilians and US and foreign senior military officers destined for executive positions. There were 213 students divided into 18 seminars of 11 or 12 students and a faculty seminar leader, and it was a one-year school. In the morning we would generally have a well-known speaker from academia, industry, government, or the international community. The subjects would range over a wide variety of international relations, economics, politics, military affairs, national issues, and current events. The groups would all return to their seminar rooms and discuss the speaker's subject in detail for the rest of the morning. The speaker would participate in one of the seminar discussions. The afternoon would be spent on seminar tasks with one of us making a presentation in our area of expertise or on an assigned project. We finished the day on the athletic field in competitive sports. Our seminar played soccer and won all of our games until the final playoff. There were several field trips. We enjoyed a very active social life. Several activities included the children, and Sabra and I made many more friends.

We were offered an opportunity to enroll in the Auburn University MBA course, which had classes three nights a week, using our academic facilities. I enrolled and enjoyed it tremendously. It was perfect timing. All of us had significant leadership and management experience and the formal management courses soon became very valuable to our careers. Both courses proved to be very relevant in all future assignments.

I had to prepare a formal research paper for the Air War College, with footnotes and bibliography, and also a thesis to Auburn to earn my MBA. In my AWC thesis I described the conceptual basis for Tactical Aircraft in 1980 and the capabilities assuming that our digital avionics initiative was successful. I postulated the significant increase in combat effectiveness through increased target kills, (via guided munitions), decreased losses (via improved situation awareness), and less collateral damage (precision guided munitions require smaller warheads for target kill). My MBA thesis applied the Cottier and Levy marketing approach of satisfying consumer needs to the design of combat aircraft and derived the military consumer needs to destroy tactical targets in a combat environment in a cost effective manner. I followed our mission area analysis methodology from the target to the aircraft and from the aircraft displays and controls to the human brain and senses. Both papers further developed and refined the ideas that our digital avionics team had failed to document, and I used both papers to recruit believers within the USAF and DOD.

My combat buddies and peer group thought it was pie in the sky. Neat analysis, but complex, expensive, and unreliable. Better to spend that money making lots of simpler fighters and continue to rely upon better trained and bolder fighter pilots, always engaging the enemy above their own territory. The Warsaw Pact conventional war fighting aircraft outnumbered NATO aircraft by a significant multiple, and their surface-to-air missiles (SAMs) were forcing both our tactical and strategic aircraft into low-level penetrations. I continued to believe and articulate that integrated digital avionics and guided munitions had to be our future* Few in the Air War College supported my views. The AWC experience was very rewarding, and going into academic, athletic and leadership competition with my peer group helped me to recognize my strengths and weaknesses.

Our whole family enjoyed the tour. We had more family time together than we'd ever had and really enjoyed it because our children were approaching their teen years and were more fun. Sabra and I enjoyed a much more active social life and everybody loved and admired Sabra. She was no longer bashful and had grown more beautiful as she grew older. She was a gracious hostess and enjoyed being with the cream of our peer group. Her name always inspired conversation at our numerous social gatherings.

I had loved her name the first time I heard it ... Sabra. We learned more about its origins throughout our life. The earliest reference, to the best of our knowledge, came from the 13th century Golden Legends—a work recounting the exploits of holy men and women of the past, with much emphasis on miracles—that tells of St. George and the dragon. As the story goes, a dragon that lived in a lake in Libya used to eat people who lived nearby. The townspeople asked St. George for help. He arrived in time to slay the dragon just as it was about to devour the princess, "Sabra." St. George is the patron saint of England and the Feast of St. George is celebrated today. He is believed to have suffered martyrdom in the 3rd century, so that places the name of Sabra as a princess in Libya at that time. An Israeli fighter pilot, whom we met at a party, waxed eloquently over Sabra and her name, and told us that a native-born Israeli girl is called a Sabra, which may be considered as a little princess. He was amazed that she had been christened as Sabra before 1947 when Israel became a nation. An Egyptian fighter pilot at the same party reminded him that Sabra was also the name of a cactus in the desert. The Israeli pilot replied proudly, "Yes, and that cactus is prickly on the outside and very soft on the inside." Sabra chose to consider it a compliment. Thus, all of our evidence points to the Middle East as the origin of her name. There are a number of "Sabras" in her family tree including her mother, two nieces, a granddaughter and two grandnieces. The seven Sabras in our life are shown in the photograph below. None of them were cacti-like, but all are princesses!

The Seven Sabras

As we had advanced in rank, we found ourselves socializing with younger groups where we were responsible to set the example. Our AWC group were the people whom we would work with for the rest of our career, and we formed deep friendships and mutual trust that would help resolve future USAF problems. The AWC and MBA program greatly expanded the breadth of my approach to life in a most positive way.

We received our next assignment to the headquarters USAF in the Research and Development deputate. I was assigned to the aeronautical system division as the branch chief for all of the research and development technology programs for USAF aircraft. I would be in the Pentagon for the next four years. We graduated from AWC and I stayed on for an extra month to complete my MBA. I only had to check the AWC bulletin board every morning and go off to class. Right before I finished my MBA, I checked the bulletin board and the Colonels promotion list was posted. I WAS ON THE LIST! I had made Colonel below the zone again. Everyone from the AWC was on leave because school was out, so nobody had informed me. I rushed home to tell my family and we called all of our family and friends. Most of our military friends already knew it because the promotion list had been out for several days. I had served less than two years as a Lt/Colonel, one year of which I was in school, and made BIRD COLONEL! I made several weekend trips to Washington D. C. to look for housing. The prices were outrageous. We would have a very hard time continuing our savings program, even on a Colonel's salary.

The Pentagon Plus: 1971-1975

We purchased a brand new four-bedroom home on Little Hunting Creek in Mount Vernon, right across the creek from George Washington's land boundary. I would drive to the Pentagon up Mount Vernon Parkway, and we could jog and ride our bicycles on the parallel bike trail from George Washington's Home (1.5 miles south) up to Alexandria, Virginia (9 miles north). I paid $145,000, a huge increase from the larger four-bedroom home in Huber Heights that Sabra bought for $22,500 just six years earlier. It was well located, and we had our children in Catholic schools again. We had sold our home in Huber Heights for $33,000 to make our down payment of 5% on our new home. We always used our VA loan to obtain maximum leverage.

The Pentagon was frustrating: a lot of meaningless meetings, continual rewriting of correspondence as you coordinated your paper product with everybody and their brother, continual pressure from Congress as we worked our budgets, and long work hours with many working weekends. The reward was that we were working with the best minds in the USAF from the Generals down to the Majors. We also had occasional opportunities to influence the future of the USAF. If I had plotted my assignments, and I never had, I would have given me my current job. I worked hard to implement my AWC and MBA thesis. One day I was asked to update and coordinate the JCS Long Term Technology Plan. The harried Army Major trying to coordinate it could not find anyone to pay any attention to it, much less update it. Everyone, including our Congress was frantically working on the next two years' budgets. The long-term does not exist in Washington D. C. When I offered to revise and update the USAF portion, the Army Major almost fainted. I worked with our USAF Systems Command and used their inputs from the USAF Laboratories, the Space System Division, and adapted my AWC and MBA work into the desired JCS format, creating our new long-term goals and objectives for the USAF. It took me about two weeks of overtime to complete. Since the document was widely ignored, it was easy to coordinate and the Army Major did all of the Army, Navy, and JCS coordination. It was meaningless but it provided a baseline to build from. I began to refer to it frequently. Now I had to obtain seed money for our USAF Laboratories to build the required technology base. I had good contacts in the Labs and consulted frequently with Jerry Pasek, now Major Bill Crabtree and others to fund the most relevant technologies.

The Assistant Secretary of Defense, Dr. John Foster, was a brilliant scientist and a long-range thinker. In 1972 he requested the R&D Assistant Secretaries of the USAF, Army, and Navy to submit new initiatives to demonstrate how our technology could be exploited in future weapon systems. Our three-star General selected Major Del Jacobs and me from our Requirements group to propose candidates. We selected the Lightweight Fighter (which became the F-16), the Advanced Manned Short Takeoff and Landing transport (which struggled and died), the Fly-by-Wire flight control system (which eventually went into most military and commercial aircraft), and the Digital Avionics Information System (Our Lab team's idea, which evolved to be applied in most military aircraft). Two of the four new initiatives were in my branch and all exploited the technology that came from my branch. All four were funded, and the Avionics Lab formed a new Division concentrating on integrating multiple avionics functions and optimizing information outputs. Our mission was funded, had a talented organization executing our plan, and our idea was off and running.

I was appointed acting Chief of the Aeronautical Systems Division in the R&D deputate, which was the staff support for the development of all USAF aircraft, subsystems and technology. It was a General Officer's slot. Our biggest job was preparing the congressional testimony for our Generals. It had to be consistent with all of our prior testimony, and we had to respond to any related congressional queries within 72 hours. All congressional testimony had to be thoroughly coordinated within the Air Staff. No mistakes were tolerated, and a single mistake was the end of your career. After a short time I became aware of how Congress played the game while we took the blame. We were flying old aircraft and badly needed to replace our fighter fleet, which had been flying many hours now in combat within Vietnam. We recommended that we terminate production of the F-111 and A-7D and use those production dollars to start production on the A-10 and F-15. The Secretary of Defense agreed and included these changes in our production budget. When it came back from Congress, we had some money to start production on the new aircraft, but they had not stopped production—only greatly reduced the production rate—for the F-l 11 and A-7D, resulting in a very large cost increase in each aircraft. The press made a big issue of it, and we were blamed for the gross overrun. The Speaker of the House made the changes and we received the blame. It was all about jobs in Texas, so we ended up continuing to produce old aircraft that were less effective than our new aircraft at a much larger cost. Our Congress used the defense budget like a pork barrel. I lost a lot of respect for our Congress during my Pentagon tour.

The government of Japan requested assistance in structuring an Air Defense system for Japan, primarily against attack from North Korea. I led a team of four people and we spent a week in Tokyo as guests of the Japanese Government. I was the RADAR expert, due to my experience on demonstrating the next generation RADAR. It was a luxurious stay, and they were a very gracious host. I believe we turned them around on their intended approach, and we received a very complimentary letter through our Department of State down through our chain of command.

Toward the end of my Pentagon Tour, after our Congress cut off the funding for the Vietnam war and we deserted South Vietnam, we received our POWs back from Hanoi. A very large number of them were F-105 pilots whom I had known. Several of them were listed as MIA because the Vietnamese never listed them as POWs. The Red River Rats threw an enormous party in Las Vegas to welcome them home, and we converged on Las Vegas at our own expense to enjoy our long-sought reunion with our comrades. It was the highlight of my Pentagon tour, and my bosses allowed me to go. I think Ross Perot, who was instrumental in bringing the political pressure to bear, subsidized our giant party. He will always be a hero to all of our families and particularly our returned POWs.

I had one more year to go on my Pentagon tour when I was called in by my three-star general, General Evans. He had nominated me to be one of six USAF colonels to be in a "Think Tank" reporting to the four-star general, General George Brown, who was the head of the whole USAF. I was honored. All of us who were nominated were interviewed by General Brown. I was one of the six selected. General Brown was a long-term planner and thinker. He had come up through his career to the top and, upon arrival, found himself surrounded by the brightest of his peer group. All of them were busy as bees, putting out short-term fires and inundated with speaking engagements and social requirements. He gave us our mission: to first interview several men that he considered long-term thinkers and then to develop, as best we could, future strategies and concepts for the USAF. He assigned us offices in the Air War College at Maxwell AFB, Alabama, where the USAF Air University Library was located. We were given full security clearances and a generous travel budget. Our ideas were to be submitted directly to him without intermediary interference. He told us that he wanted the luxury of knowing that he had six highly recommended people thinking about the long term for the USAF. I was enthusiastic. This would count as my fourth year in the Pentagon, so I escaped with only three years in the actual building.

The Six-Man Group: 1974-1975

Sabra was delighted to have me back again in the family. We rented our home in Mount Vernon to a USAF Colonel with a Ph.D. in Physics and moved to Maxwell AFB Alabama, where we were given General Officers Quarters—large brick two-story buildings with twelve foot ceilings built by the WPA in 1930. They were magnificent. We met the other five Colonels and their wives: Colonel Pete Piotroski had been the Fighter Wing Commander at Aviano AB Italy; Colonel Stuart Bowen had been in Plans for the Air Force Systems Command; Colonel Bob Reed had been HQ USAF in the Pentagon in Plans (Concepts and Doctrine); Colonel "Moon" Mullins was Director of Operations in a Fighter Wing in the Tactical Air Command, and Colonel Len Siegert was a staff officer in the Strategic Air Command. Colonel Len Siegert was our only Navigator, and he had not brought his family because it was only a one-year tour. All the rest of us were fighter pilots and had brought our families.

We all had met in General Brown's Office and were notified that we had no leader but could work as individuals, partners or a group. This was a wise decision to avoid chain-of- command problems in our think tank. We all had a lot in common and socialized readily. After we all had moved to Maxwell AFB, our families met and they were all compatible.

We started our travels to interview the people whom General Brown wanted us to listen to: Werner Von Braun, the German rocket designer who now worked for NASA; the Chief of Staff of The Royal Air Force in England; General Dixon, the Commander of the USAF Tactical Air Command, who had served in a similar think tank when he was a Colonel; Lt/General Evans HQ USAF DCS R&D, who had nominated me; the Secretary of the USAF; and a well known scientist and writer whose name escapes me. All were very bright and revealed their thoughts on the future international and political military environment. It was very stimulating, and I began thinking about the impact of space technology on warfare—a Pandora's box to which I had very briefly been exposed. We all had been reading a great deal about airpower, lessons learned in war, and philosophy exploiting the Air University Library. During one of our bull sessions, I had just reviewed "The Principles of War" and remarked that so far the impact of nuclear weapons had reduced grand strategies to deterrence. Our Strategy of Mutual Assured Destruction (MAD) was a deterrent strategy. I stated that as far as I could tell there were no "Principals of Deterrence." The group lit up like a Christmas tree and Bob Reed suggested that we start investigating

that as a potential project. The group agreed and we began studying, discussing, and finally writing and publishing "Principles of Deterrence" in the Air University Magazine, which was widely read throughout the USAF. All six of us were listed as authors but I had contributed very little after our initial discussions. I had taken off on my space investigations and had made several trips to our Space System Division in California. I had worked with them before when I had the Satellite Interceptor Program. I wrote the first proposal for General Brown. I was still not ready to suggest warfare in space because I believed that we had a clear technology lead in the potential to use space for reconnaissance, secure communications, surveillance, and navigation. These technologies would provide significant advantages to our battlefield before, during and after a war. However, I believed that a midcourse ballistic missile interceptor was a necessary ingredient to support our current strategy of deterrence. The only way that a nuclear power could win a nuclear war was to decapitate the opponent by a surprise attack; checkmate, so to speak. If the President and his successors could be killed, we would be constitutionally neutered. If we developed and tested such a capability, that might be sufficient to deter the attempt. The rest of the group were not comfortable with my paper but I sent it in. My next paper was going to address the issue of the value of our space capabilities in future warfare ... but lightning struck. Our boss, General Brown, was promoted to Chairman of the Joint Chiefs of Staff, the senior General of all of our U. S. Armed Forces.

The new Chief of Staff for the USAF, replacing General Brown, was General David Jones. He discovered he had six colonels in Maxwell AFB reporting to him, tasked only with "thinking." He promptly changed our mission to look for near term opportunities to reduce the operations and maintenance expenses for the USAF. We would serve for six more months and receive "good" assignments according to his executive officer. We began looking for ways to reduce budgets and General Jones requested that one or two of us look into combining the Air Force System Command and the Air Force Logistics Command. Stu Bowen and I were best qualified to do this and we went to work. The other four prepared proposals to save money in the other commands. General Jones requested that we brief the Vice Commanders of the impacted Commands at Maxwell AFB and then send him the completed studies and an overall briefing. Bob Reed (bless his heart) volunteered to consolidate and brief all of the three-star Generals, and we would field their questions. The five three-star Generals flew down to Maxwell AFB to receive the advice of six Colonels on how to save money in their commands!!!

We were not looking forward to this visit. Bob Reed did a wonderful job of presenting our briefings. All of the three-star Generals were polite and it was clear that they understood the position that General Jones had put us in and decided to make it easy for us. They were truly "gentlemen," and I was proud that they had not taken it out on us. We breathed a sigh of relief and made plans to move to our new jobs. The final blow was that we were told to write our own fitness reports and the CSAF would sign them. There was very little to say and it probably was my worst fitness report. I never saw it again but I figured that the best job I had in my life only lasted six months. I was hoping that I would be given a command position now, since I was ripe for it.

I was assigned to Wright Patterson AFB again in HQ Air Force Logistics Command. I was the Assistant to the Deputy Chief of Staff for Acquisition Logistics, a new organization formed to assure that all new systems were designed to be available and affordable. It was the first of my "new organization" jobs, which would represent the rest of my career.

CHAPTER 15

New Organizations: 1974-1980

When a problem surfaces, one option is to form a new organization to resolve that problem. Within government, that results in using people and funds that are diverted from other organizations until your new organization obtains its own manpower and funds, which may take two years. If the problem is important to your executives, you may handpick your people. If not, you are provided the people that the other organizations are willing to lose. Regardless, there is a lot of turmoil and stress during the start-up. This organization had been started eight months earlier and the startup phase was almost completed; however, my boss was the third General and I was the second deputy to hold the top two positions. The start-up was accomplished by Major General Buckingham because he had the power to grab the right people. Now Brigadier General Rutter and I were the permanent team selected to run the organization.

Major General Buckingham and I had worked together in the Pentagon, and I had been very aggressive in emphasizing reliability and maintainability in the R&D budget actions. He was a Colonel then in the R&D Budget, and he would require my coordination on budget items, knowing I would insert some language that would require reliability. I also found allies within the logistics community to join our plot. One of the Assistant Secretaries of Defense for R&D, Dr. Jacques Gansler, was emphasizing Design For Cost for the R&D

community but only talked about production cost. I was trying to change it into Design for Life Cycle Cost, which would include operating and maintenance cost. I had several sessions with Dr. Gansler and he listened but never changed his slogan. I knew I had made my point with him. I am sure that these debates were the reason that both General Buckingham and I ended up in the USAF Logistics Command. It served me right. Now I would have the opportunity to make Design for Life Cycle Cost the USAF slogan.

Sabra and I moved into Colonels Quarters in Patterson AFB. Our children had always been in Catholic schools, and now I thought they should go into the local public high school in Fairborn, Ohio. Sabra was reluctant, but I wanted them to obtain some public school experience. It was the dumbest idea of my entire life and I still regret it. The teachers were poor and the students were undisciplined and sloppily dressed. My children made better grades but learned much less. All three of them graduated in less than four years because, in their Catholic schools, they had completed much more than their Fairborn peers. I allowed them to graduate early to get them out of that environment.

Kellie was taking Ballet lessons in downtown Dayton. Kevin wanted to do some Christmas shopping, so he dropped Kellie off and parked nearby. As he was walking away, he was stopped by a rough-looking young man, who explained he needed $20 for a prostitute. When Kevin showed him he did not have that much money on him, he took what little money Kevin had and said that Kevin would have to snatch a purse to get rest of the money. He explained that Kevin had to do the robbery because he, the crook, had just been released from prison. He implied he had a knife in his pocket and would use it on Kevin if he did not cooperate. They went back to the car and drove until they saw an elderly lady walking alone with a purse. Kevin got out of the car and headed toward the lady. At the last moment he took off running as fast as he could. The black man caught him easily and they were on a bridge over a small river. He threatened to knife him and throw his body in the river. He drove Kevin to a bar and introduced him to several shady characters. Then they went back to a shopping center, parked the car, and walked up to where people were leaving the grocery store. The man selected another elderly lady and told Kevin to follow her and take her purse when she was out of the lighted area. Fortunately, she walked toward Kevin's car. Kevin grabbed for the purse but she held on and fell down and Kevin ran to his car, jumped in and locked the door.

He had noted that the keys had been left in the car. He saw someone running toward him, started the car, and accelerated straight ahead through the snow and onto a major street, entering at a 90-degree angle, turned right, and zoomed off to freedom. He stopped at a restaurant and the manager loaned him a dime to call us. Kellie had finished her ballet class and Kevin was not there to meet her, so she had called us. We were worried to death and reported Kevin and our car as missing to the police. When Kevin called, we immediately called the police and they requested all of us to come down to the police station. The lady had reported the attempted purse snatch and Kevin admitted the crime and told his story. The crime had also been seen by a plain-clothes officer, and he was the one chasing Kevin in the parking lot. The officer had also seen and recognized the man as a well-known robber, but he was gone when he returned to the scene. The detective was disgusted, because a judge had released this man from jail on what he called "shock probation" after serving less than one year on a two-year sentence. He felt he could find him and get a longer sentence now for violating probation. The trial was about three months later. Kevin and the lady testified and the black man glowered at Kevin. The judge sentenced him to time served and he walked out of jail. We did not know the sentence until the next day and were furious. I called the detective and he was furious. I was concerned for Kevin's safety, but the detective assured me that he could tell this criminal where Kevin lived, give him instructions on how to find our home, and give him bus fare to get here and he lacked the mental capacity to find us. He was very concerned, though, for the safety of the public because that man could be dangerous. He also stated that this judge had released another criminal on his "shock probation" experiment and that criminal had raped a woman shortly after his release. I vowed to get the judge. The detective later called me and said that the judge was a very heavy drinker. The story proved to be true, and the judge was removed from the bench. His two "shock probation" experiments would have made a good newspaper story. The robber was back in jail within three months. I had more sympathy for the robber than the judge.

I inherited an outstanding group of young Captains who had master's degrees in Acquisition Logistics from our Air Force Institute of Technology and, like me, were assigned to a job that directly related to their education. I requested the job of selecting the USAF officers to go into our graduate schools in Acquisition Logistics and in return was allowed first choice from their annual graduation of brand new graduates with master's degrees in Acquisition Logistics. I selected the top performers in the various maintenance disciplines within the USAF who had volunteered for graduate degrees and then could cherry-pick the best of their graduates two years later. We placed these officers in the System Program offices that were developing the next generation USAF aircraft, space vehicles, and missiles, and they provided excellent advice to the program managers on the design of a logistics support system to optimize life-cycle cost for that weapon system. I maintained a staff of these master logisticians to provide this support to all less-than-major weapon systems and to develop life-cycle cost models using existing data systems to stay abreast of reliability and maintainability innovations. We also started a Lessons Learned in Acquisition Logistics to document the results of our good and bad ideas. I had about half military and half civil service employees and my civil service employees would be there to implement our processes into the USAF logistics institution, while our military would be transferring around the USAF winning converts. Design for Life Cycle Cost eventually became a way of life after we developed the models and could validate them using our data systems.

The organization was successful and the four-star general took us out of the headquarters staff and made us a Division. The Acquisition Logistics Division was put under the command of Lieutenant General Bryce Poe with B/General Rutter as deputy. A new B/General arrived to run the less-than-major weapon systems, and I ran the major weapon systems, which was the fourth General Officer's position that I had served in.

During our reorganization I took a week's leave and Sabra and I went to St. Louis to visit my mother and Aunt Gladys. My great aunt Eva was dying, and we drove down to Geneva, Missouri, to visit her in the nursing home. Mom wanted me to wear my full uniform so that my doting great-aunt could have one last great dote. She was 94 and had never married but had worked for the President of Monsanto Chemical Company until he finally let her retire in her late seventies. I was very proud of her and she was as positive and joyous as ever. We spent the day with her and she died three weeks later. We drove back to the base, and Tina was out in the front yard all excited. She told me that General Poe had been trying to find me and she had heard that I had made Brigadier General. So I called in, and General Rutter told me to get in my uniform and report immediately with my wife to General Poe. (Traditionally, your commander is the only

Aviation Cadet
Robert W. Kennedy, 1952

Brigadier General
Robert W. Kennedy, USAF 1977

one to tell you when you make general; not your daughter!) He was very happy to tell me and everybody on base knew because the list had come out that morning. They could not find me because we were driving back from Saint Louis. That was my second promotion where Sabra and I were among the last to know.

We threw a very nice promotion party in the officers' club with all of my family, my workers and friends around the base. We all had a great time. Sabra was a warm and gracious hostess and her bashfulness had long since disappeared.

Sabra was very popular and our four-star General Mike Rogers was fond of both of us. He was an American volunteer flying with the RAF during the battle for Britain. He was an advocate of combining the Air Force Systems Command with his command, the Air Force Logistics Command, which Stuart Bowen and I had written for General Jones. He was a very enthusiastic, cocky ex-fighter-pilot who had designed his own formal uniform with a cape, and he carried a swagger stick. He called me in informally to discuss the study on occasion and tell me fighter-pilot stories. He had appointed Sabra and me to run the annual Debutantes Ball, which consisted of a series of teas, dances, and the formal coming-out ball. Sabra selected a few other wives and they supervised the young debutantes. I watched over the young men selected to escort the debutantes. These were mostly children of the career officers and NCOs on the base who were home from their first year of college during the summer holidays. We had fun with the young people, and our daughter Kellie came out as one of our debutantes during our second Ball.

My job remained about the same. We were gradually accepted by the Air Force Systems Command, because we obviously had been adding value by reducing the life-cycle cost of our next-generation weapon systems. General Rogers and Lt./General Poe had some Pentagon support for us to take over the management of some of the weapon system acquisition programs from the Air Force Systems Command, and they were providing us with more stars: Major General Albert replaced B/General Rutter as Deputy Commander, and I was promoted to Brigadier General.

One year later General Rogers announced his retirement and Lieutenant General Poe was promoted to full General, replacing General Rogers. Major General Albert was promoted to Lieutenant General and replaced General Poe. That is a lot of promotions for one of many organizations within the Air Force Logistics Command. Our entire organization was enthusiastic, because we obviously had proven our value to the USAF, and I looked forward to greater acceptance within the acquisition community. My next assignment came suddenly via a phone call from General Poe to my home.

General Poe had returned from his courtesy trip to the Pentagon where he met with the Secretary of the Air Force and the USAF generals in Headquarters USAF as a courtesy to receive his congratulations. Two days earlier, Senator Eagleton from Missouri confronted the Chief of Staff of the USAF in the halls of Congress following his testimony. The Senator had his TV makeup on and several TV camera crews. The very senior Senator stated, (paraphrased) that the USAF should be held criminally liable for their mismanagement of the Air Force Logistics Data Management System. The next day two USAF generals were retired and General Poe showed up for his courtesy calls in the Pentagon as the new Commander of the Air Force Logistics Command. His courtesy calls turned into discussions of these charges and he came home rather upset that he had been blindsided. He called me and informed me to be in his office early the next morning. I was appointed as the new Director of his Air Force Logistics Command, Logistic Management Systems, and would be reporting directly to his very capable Vice Commander Lt/General George Rhodes, my neighbor on General Officers row.

The next morning I met with General Poe and General Rhodes and we formulated an approach. I would be the Assistant to the Commander for Logistics Management Systems and allowed my selection of the best 10 functional specialists in the command, plus a deputy who would be selected from the best data automation experts within the USAF. He made it very clear that I would have this job until I had a recovery plan for our Logistics Management Systems approved by Congress. He also told me to always be totally truthful as to our status and he would stand by me with all the support he could muster. As it turned out, this was a vital element because a lot of blood had already been spilled and both sides were eager to retaliate. The background was littered with turmoil.

During the early 1960s AFLC were widely considered the leaders in developing and operating state-of-the-art data management systems within the Department of Defense. The commercial companies were ahead of us, because U.S. Government Data Automation were constrained by a law sponsored by Congressman Brook of Texas. The rationale for this law was that IBM enjoyed a virtual monopoly on providing data automation equipment within DOD and Industry. The Brook's Bill was structured to eliminate IBM's business advantage. Congressman Brooks passed a law that required that all future U.S. government Data Automation procurements had to be "fully competitive" so that all companies could compete on an equal basis. Most users were already equipped with the superb IBM equipment, and IBM guaranteed that they would preserve upward compatibility for the software and ancillary devices for the next two generations of new technology equipment. This provided a strong incentive to stay with IBM when you needed the next generation technology; i.e., you could reuse all ancillary equipment and your software. Each computer company had their own unique devises and many used different languages. Therefore, in 1965 AFLC decided to totally satisfy the Brooks Bill by running a fully competitive program to procure all new modern technology Data Automation equipment and simultaneously redesign all of its software programs into better-integrated packages. They were operating on IBM 7080 computer hardware in machine language (autocoder) with more than 400 separate functional programs. There was no common dictionary for all 400 programs, and the update intervals and "as of dates were random. The programs had to be printed out on paper and functional or weapon-system users had to refer to many printouts to obtain needed data. We were locked into IBM for two more generations, which was against the law, unless we fully competed with all hardware and software, which is very risky. The technology was improving fast, and our equipment and programs were not effectively satisfying our information needs. In 1965 AFLC started its fully competitive procurement for the Advanced Logistics System (ALS), which included all of the hardware and software. Cubic won the contract.

When they turned the key to start their new Data Automation Equipment, using their brand new integrated Logistics Management Systems, it did not work. Hundreds of millions of dollars and 10 years later in 1975, Congress canceled the ALS program after years of severe criticism. Now we were stuck with our very old and obsolete hardware and software, which could not continue to support our wholesale and foreign military sales logistics management systems. AFLC started a recovery program that grew even more controversial, exploding in 1978. Two Generals were fired and AFLC was in an emotional dispute with HQ USAF and the Secretary of the USAF. The Senator's angry accusation of criminal liability on the evening news elevated it to a scandal worthy of national news. Now I had to resolve it.

My deputy was the best Data Automation Officer in the entire USAF, Colonel Rick Lejk. This highly qualified state-of-the art expert and experienced manager now was going to help me patch our obsolete hardware and software and procure a modern system without violating the Brooks Bill. It would require an Irishman and a Poland-Person (not a Polack) to resolve it. Per General Poe's direction, we went to Washington to talk to all of injured parties who felt we had ruined their reputation with the Congress. They all started with, "I know you did not cause this problem but—" and we then were warned about all of the emotion involved and what not to do or say. It was obvious that positive communication would be difficult, and I remembered General Poe's very wise advice: Be totally honest on everything.

We had an open brainstorming session with our ten new employees, which lasted several days, and formulated a tentative plan of action.

We would not take over the AFLC Data Automation group. They had been cut from 1800 programmer/analyst to 1200, and their morale was very low. We would take this problem off of them, which would help improve their morale. Managing 1200 programmer/ analysts would severely disrupt our primary task. It was now our problem and we were responsible to fix it. We had to convert all existing 400 logistics management systems from autocoder language into a higher order language to be able to run them on modern computers, because our computers were not going to last long enough to procure new computers. That required 700 programmer man-years. We formed five functional groups and would appoint a top civilian project manager to prepare the functional requirements for those five newly designed logistics management systems that would replace our 400 poorly documented systems. Each project manager would have a programmer analyst as deputy manager to manage the design and coding of the software. This placed the responsibility for the design on the functional users of the product. Our group would be responsible for maintaining a common dictionary, "as of" dates, and documentation and coding standards. These five logistics management systems would be ready for operation by the time that we could procure our new data automation computer system. The IBM 7080s would have to last until we could fully compete for a new computer system. There were only 20 of these computers left operating in the world, and we had made arrangements to buy them to use for spare parts when they were retired. IBM had already notified us that the IBM 7080 transistor was no longer produced. The risk was that the IBM 7080s would fail before we converted our autocoder software into a higher order language. Then our only solution would be to purchase IBM 360/65s or IBM 370-168s with emulators that made them look like IBM 7080s. That was the excellent and low-risk solution that caused all of the turmoil with the Washington people who claimed it violated the Brooks Bill. I thought I might be able to sell it this time because we were going to convert our software and start a fully competitive procurement for modern computer equipment. We would only use these forbidden equipments as a backup in the event our IBM 7080s failed. If the IBM 7080s failed, our USAF Wholesale and Foreign Military Sales Logistics would have no data systems support!! Those are unacceptable consequences. I certainly was not going into this high-risk program without a backup solution. With all of the constraints placed on us, there was very little else that we could do.

I was required to provide a 20-minute briefing on our progress to the HQ USAF Staff and to the Assistant Secretary of the Air Force for Financial Management every 30 days. We were ready for our first briefing. The 20-minute briefing to the Air Staff took five hours. Everybody nitpicked me to death, but we only changed the wording; not our approach. The 20-minute briefing to Secretary Hewitt took four hours, and at one time he was standing on his feet and shouting at me with tears in his eyes. Secretary Hewitt was a political appointee, and he did not want to believe how bad our situation was. When my briefing was finally over, he called General Poe and wanted me fired. General Poe told him he had reviewed the briefing, he would not fire me, and that all briefings were going to be rigorously honest. When I returned to Dayton, he told me about Secretary Hewitt's reaction and I told him about my five- and then four-hour ordeal. It was like that for 12 months and they would not approve my briefing, which by now was our Plan, to go up to the Office of the Assistant Secretary of Defense for Logistics and thence to the appropriate committee in Congress for approval.

Meanwhile, my program was being investigated by the USAF Inspector General, The OSD IG, the House Appropriations Committee staff, and the House Government Operations staff. I cooperated fully with all of these investigations because most of the emotion was caused by very poor communication between these staffs. The House Government Operations committee was responsible for oversight of Government Data Automation, and they were the ones who would approve or disapprove our plan. When their two staffers finished going through all of our records, they stopped by my office to talk to Colonel Lejk and me. They knew Colonel Lejk by reputation and had additional questions. I could tell they still were uneasy because all of the bad press had led them to be skeptical. I offered to finance two independent consultants, to be selected by them, to come down and do a more thorough examination of our requirements, processes, and detailed plans, or whatever they wanted. I only requested that they allow Colonel Lejk and me to sit in when they gave their report. They were delighted. I quickly informed General Rhodes and General Poe and they agreed, hoping as I did that once and for all this would resolve, or at least better define, our credibility problem. Two excellent consultants, well known by reputation to Colonel Lejk, spent a week examining all of our data and equipment. Basically they validated the serious potential impacts of our hardware and software problems on our ability to support a wartime surge and recommended a more aggressive equipment plan—which the HQ USAF and Assistant Secretary of the Air Force had denied to AFLC. The two staffers were convinced but would not endorse the more aggressive equipment recommend-dation because it would seem to conflict with Senator Eagleston's criminal liability statement to the press. I still had to go through our whole chain of command to obtain formal USAF, SAF, and OSD approval of our plan, but we knew that the House Government Operations Committee would approve it when it was received. It was two more months of 20-minute briefings that lasted two hours before we finally got the plan to Congress. It was approved. Mission accomplished. I was free to go. General Poe awarded me the Distinguished Service Medal before I departed.

Right before I left, I went into the office. Colonel Lejk always was there first because he felt he should be. I had previously slipped in several times at 6 a.m. just to upset him—we were always doing one-up tricks to each other. This morning Colonel Lejk was not to be seen. I went into my office and there with his feet up on my desk was Colonel Lejk! He asked me if I had read the morning paper and I replied that I had. "Well then, you know that we now have a Polish Pope and now I am the boss." We had had a very stressful 15 months together but never lost our sense of humor. We still correspond.

Our family life during these three different jobs at AFLC covered the last few years of high school and the beginning of college for our children and the job of mother, hostess, and wife for Sabra. She excelled in all three and kept me in good shape and humor. There were always promotion parties, visitors to entertain, and community affairs with our local political and industrial leaders, and we truly enjoyed all of them. We had cordial relations with our neighbors on General's Row and thoroughly enjoyed our life on the base.

I had had three start-up jobs within five years, but when I was home with my family, the stresses were forgotten or deferred. I had always kept myself in shape by jogging four miles, four times per week and playing handball, tennis or racquetball. When I made General, General Jones (the Chief of Staff of the USAF) sent all 25 of us on the promotion list to an alcohol-rehab center for three days for an unstructured orientation. I thought it was ridiculous and almost insulting. After those three days, the next time I went to the officers club I informed the club manager to change my drink of choice from a light scotch and soda to a club soda with a twist. The rehab was an awesome learning experience, and I realized that our busy social life involved scotch at least three to four times per week. I had seen the results of social drinking that had expanded into alcoholism ... and it scared me.

Kellie was enrolled at Ohio University with a ballet scholarship. Kevin studied Computer Science at the Ohio State University. When Kevin was about seven years old, he had said that when he grew up he was going to Ohio State University and live in Ohio. So far, he has never changed that plan! Tina was in Wright State University but lived with several girl friends in town. She was not enthusiastic about college and was in love with a neighbor's son who had dropped out of high school and joined the US Navy.

We received our orders. The General in charge of Operations for the USAF had formed a new Directorate within Operations Hq USAF, and I had been selected to be the Director. This was my fifth new organization job, and it became clear that I was destined to be a problem-solver in staff positions forever, never to command.

The Pentagon Again

Electromagnetic Combat: 1979-1980

Sabra, Kellie, and I moved back to our home on Creek Drive in Alexandria, which we had rented while we were in Dayton. My job was very sensitive, and I had to obtain permission to live outside of General Officers Quarters on Boiling AFB. Kevin was in college, and Tina stayed in Dayton with her roommates. Several of our old neighbors were still there and welcomed us back. Once again, I would commute to work on beautiful Mount Vernon Parkway.

I was the Director of Electromagnetic Combat, a new job created by the Deputy for Operations for HQ USAF. The job was to better focus Electronic Warfare, Defense Suppression, Command and Control Communications Countermeasures and be the Focal point for the USAF Signals Intelligence Command, a component of the National Security Administration (NSA). The objective was to provide ready access of all of these capabilities and information resources directly to our battle commanders in our war rooms throughout the world. We could control most of our capabilities from our war rooms now, but much of the electronic intelligence went to other locations for analysis before being presented to our battle commander by his intelligence staff. This time lag needed to be shortened and the capabilities, intelligence, and real-time situation displays be made available to the battle commander and his staff in an optimum information format. This presented many security problems and issues, because much of the data came from very secure national resources. I was enthusiastic because this was similar to the job of providing a single-seat fighter pilot with all of the relevant information available from all of his on-board and off-board sensors, providing total situation awareness to facilitate rapid decisions. In this case it was not an engineering design problem but required breaking down longstanding turf battles and security barriers. The design problem was easier.

This Directorate was the idea of the three-star General Deputy for Operations who was reassigned just before I arrived. The new three-star, General Gabriel, thus inherited a new Directorate with a director whom he had not selected. It was clear that he had little interest in this initiative, and he delegated it to his two-star deputy, General O'Malley, who was very familiar with all of the elements. The new Chief of Staff for the USAF was Four-Star General Lew Allen, who had been the head of NSA. Admiral Inman had replaced General Allen at NSA, and both of them encouraged me. Within the first month I learned that there was a lack of enthusiasm for this initiative within both Operations and Intelligence in the HQ USAF. Major General Larson was an Air War College classmate and commanded the Signals Intelligence Command. He was enthusiastic about being better integrated into the War Rooms. The Electronic Warfare (EW) mafia were very enthusiastic and considered my position as being primarily EW. Thus the past and present commanders of NSA, the commander of the USAF Signal Intelligence Command, the EW cult and I were positive. Operations were lukewarm but had to support it because their former boss had created the idea and Intelligence was supporting me but only on the surface. Before I even got started, the Defense Suppression people opted out of my organization over my objections. This reinforced the perception that our office was going to be an Electronic Warfare advocate. Most of my assigned staff were EW officers and my emphasis had to be the destruction of the enemies' Command and Control Communications nodes for their Air Defense networks early, so that our air forces could penetrate their very formidable Air Defenses. Our EW capabilities had not kept up with their many new excellent Surface-to-Air Missiles(SAMs) and their interceptors allegedly possessed RADARs that could look down and see our aircraft flying low, trying to hide in the RADAR's ground clutter. We had to know and map all of their air-defense sensor and communications networks and remove the nodes in the first few hours of the war, and that requires real-time intelligence and node suppression. Then you can take out their SAMs, airfields, and ground forces.

That was how I understood my vaguely defined mission.

General O'Malley recommended that I visit our command centers in Europe and the Pacific and discuss their growth requirements and opinions within my vaguely defined area of responsibility. The visit to the Pacific was very good and I learned a lot in Hawai, Japan, and Korea. All were very cooperative and mildly enthusiastic. In Europe I had very little cooperation from the intelligence officers and some uneasiness from the operations officers. I visited several of our SIGINT sights in both theatres and was received enthusiastically. General Larson's troops were the highlight of both the Pacific and European visits.

Upon my return, I learned that the Major General in charge of Intelligence had been receiving reports from his intelligence officers wherever I had visited. After my successful Pacific trip he had sent a message via back channels (normally untraceable) to the senior Intelligence officer in Europe, urging minimum cooperation with me. I was not allowed to see the message or discuss it, but that Major General retired within a week. General Allen and I were the only ones (that I know of) who were made aware of the message. This was a gross violation of ethics—presenting a cooperative attitude to me while trying to undermine my approach behind my back. I badly needed support from our Intelligence community because they were the group that was to experience the greatest change. The loss of the Defense Suppression and Intelligence communities in the first few months significantly reduced our probability of success. I had failed to build the team I needed to achieve what I regarded as success. It was a political loss on my part, but I had never envisioned that I would require political support from my fellow General Officers. All of my prior experience was with Generals who clearly expressed their view and their opinion of my view. Then we worked out a compromise based on the USAF objectives. Perhaps some of our Generals in Washington were picking up the tactics of our Congress. My observation was that there was a world inside the beltway and the real world outside of the beltway. I was disappointed, but I had underestimated the complexities of the problem.

My boundaries had been redefined for me. I chose to exploit my support within the EW and Signals Intelligence communities and facilitate the linking of our real-time sensitive intelligence data into our war rooms. I believed the operational commanders would do the rest.

I was asked to speak at the EW Old Crows annual symposium and given a Scenario to attack several highly defended targets in Yugoslavia. My USN counterpart was given the same scenario. The targets were surrounded by all of the most modern Soviet SAMs, and Anti Aircraft Artillery. The Old Crows knew that both myself and my USN counterpart knew it would be a very high-loss-rate mission, regardless of what mixture of aircraft and tactics were used. They knew that I had a good electronics background and perhaps hoped I would admit it was too tough to do and advocate some of the more sophisticated electronic countermeasures currently being proposed be funded as soon as possible. My USN counterpart and I did not coordinate our speeches but neither of us took the bait. He stated that he was not in R&D but that any attack would have high losses. I described my attack using a very low-level attack across the Adriatic Sea to destroy three command centers that were the search sensors and communications nodes that detected penetrating enemy aircraft and provided location cues to their airborne interceptors and SAMs. Our aircraft came in very low at night and proceeded directly to destroy these three nodes. We lost three of the 12 F-111s. The follow-on attack was 10 minutes later, again at low level using 24 attack fighters to hit the three primary targets and 12 Wild Weasels to suppress or kill the SAMs. The SAMs could not be cued by the destroyed search RADARs and the communications links to the SAM sites were destroyed, so their commanders were cut out of the loop. Those SAMs that turned on their tracking RADARs were promptly silenced by RADAR-seeking missiles from the Wild Weasels. The self-contained ECM was not turned on because it could be detected and used by the SAM sites to point their tracking RADARs. None of our attack or Wild Weasel aircraft were lost. They had given me a credible scenario and I had introduced industry to the new concept of node suppression. They did not complain about my failure to use ECM, and it gave them food for thought.

I had failed to stimulate the operations or Intelligence communities. We did start programs to integrate some of the realtime intelligence into our War Rooms, but I wondered if they would last after General Allen retired. It took our Digital Avionics Integrated System 10 to 15 years to gain broad support. This one would take at least another 10 to 15 years.

We had periodic nuclear attack exercises in the Pentagon and I played the role of the Chief of Staff of the USAF. It was supposed to be for the entire government within the Washington Area, but the only agency that showed up was the Department of Defense. The other 26 federal agencies ignored the exercise. Only a token number of senior staff officers participated and then prepared reports with lessons learned and recommendations. We would be alerted, then rushed to the helicopter pad outside the Pentagon and flown to a secret underground location. This time the enemy missiles were impacting on the U.S. as we entered the command post. We then experienced the next 30 days time-scaled down to three calendar days. It was a sober exercise and I wrote a thorough report for General Allen, the real CSAF, closing with the recommendation that his sleeping quarters had a cot that was much too small for him. (We both were 6 feet 4 inches tall.) There were other more serious recommendations, which would impact my life later.

Our nation was in the pits. President Jimmy Carter had pulled the rug out from under our only really good ally in the Middle East other than Israel. The Shah of Iran had been a strong supporter of the U.S. for a long time. We had sold him many of his weapon systems and trained many of his military officers. Sabra and I had attended parties in the Iranian embassy and socialized with several of his Air Force officers. The Shah had cancer, and our country apparently talked him into leaving Iran before the radical religious leader, Khomeni, who had been in exile in France, returned to Iran. Khomeni was greeted by his radical supporters and took control of Iran without meaningful resistance. All of the Iranians whom we knew in their Washington embassy were recalled home and rumors of assassinations began. Then Khomeni's young radicals overran our embassy in Iran and took our people as hostages. Major General Phil Gast was the commander of my F-105 Squadron in Thailand and he was one of the hostages. It was a blunder that will never be forgotten. Just like Vietnam, we had undercut another good ally and lives were being lost because of our ignorance in world affairs and the weakness of our National Will. In America interest rates soared above 15 percent, oil prices increased, and our morale was plummeting. As a relatively minor side issue, President Carter canceled the B-1A program, which signaled another sign of weakness to the Soviets. This President, who was very honest and moral, had really blown it.

I submitted my retirement papers and retired with 30 years of service. I had been assured by General Gabriel that I would receive my second star the following year, but I was ready to leave the Pentagon. I had served in 10 different commands; Korea, Germany and Thailand; in nine different career-specialty codes and was a Command Pilot. Unlike most officers, I had a very diverse career. It had been extremely rewarding, and Sabra and I made many lifelong friends. Sabra did not want me to retire because she truly enjoyed the USAF, but it was clear to me that I probably would be in the Pentagon for my remaining five years. I was tired of 60-hour work weeks, working inside of a vault with no windows. I was replaced by B/General Bill Kirk, another Air War College classmate, who made four stars later. The directorate of Electromagnetic Combat was dissolved four years later.

CHAPTER 16

Transition to Civilian Life

The morning after our retirement party, we slept in. I was on leave for the next 90 days. I had lost leave every year, because there is no way that a senior USAF officer can find the time to take the 30-days-a-year leave. Every year I would lose about 20 days leave and be cut down to the maximum allowable 90 days of accumulated leave.

We had a wedding to plan. Our youngest daughter, Tina, was marrying her high school boyfriend, Brian Hicks. He had been in the US Navy for one year and looked more trim. Both were 18 years old. She had stayed behind in Dayton, dropped out of college, and wanted to be married in Pensacola, Florida. We went down to Pensacola, Sabra made all of the arrangements, and they were married in the Catholic Church across the street from Sabra's mother's house. We had the reception in the Garden Center two blocks away and everyone had a good time. By now we had 39 nephews and nieces, as Sabra's brother and six sisters worked hard to populate Pensacola. Tina loved all of them and stayed in touch. We had spent many vacations on Pensacola Beach and of course they would come and play with us.

Tina's Wedding

Kellie **Kevin** **Tina**

Kevin was still at Ohio State University and Kellie was living with us in Alexandria, Virginia, on Creek Drive. Kevin completed three years at Ohio State University and went into restaurant management. Both Kellie and Kevin had gone to three different high schools—Fort Hunt in Virginia, Montgomery Catholic in Alabama, and Fairborn Baker in Ohio. Kellie had completed two years of her ballet scholarship at Ohio University before deciding she did not want to be a dancer. She moved in with us in Virginia and enrolled in Northern Virginia Community College. We were glad to have one of our children living at home, still trying to find her place in life. We had always enjoyed the friends, both boys and girls, she brought home to meet us. We always wanted our children to bring their friends home and would allow each of them to bring one friend with us when we went on vacation in Pensacola, Florida, or with my sister's family in the northern peninsula in Michigan. Now I only had one child left to provide lots of guidance to, while Sabra rolled her eyes and smiled wisely.

I would not seek employment while I was on terminal leave. There are very severe restrictions on retired military officers, but not ex-Congressmen. I had my retirement physical and they found the usual problems—hearing loss (all pilots and artillery officers), my knee from being forced into the small box, and several other joints with arthritis due to my broken arms, ankles, and collarbones. The week before my physical the Washington Post had a front-page article that revealed that retired General Officers were usually retired with a disability rate greater than 50 percent. Generals all serve 30 to 35 years, a very active and vigorous life. The politically correct military reacted as usual. The next week I learned that I was rated less than 10 percent disabled. I accepted it and never appealed the rating. I was very tired of Washington's anti-military bias leftover from the Vietnam War.

On the last day of our leave, I received a phone call from a retired military friend who asked me to come in and talk to Jacques Gansler, the former Assistant Secretary of Defense, who now ran the Washington office for The Analytical Sciences Corporation (TASC). We had debated privately when I had my R&D job in the Pentagon back in 1971-1974. I showed up the next morning. He wanted to hire me to lead his studies for the President's National Security staff. We reached an agreement that I would come as a consultant for as long as the two studies required and give us both a chance to see how it worked out. Sabra and I were flabbergasted when we sat down and figured out how much we would make with my retired pay and my very generous consulting fee. We had saved and invested our money so that we could send our children to college and own homes, but now we had a very nice nest egg built up and were approaching declining expenses and my new career would put us in a large surplus on cash flow. Then I figured my Federal Income Tax and I was in the 70-percent bracket. This was ridiculous that the wasteful government would take 70 percent out of every extra dollar I earned. President Carter had given us an out with his shattered economy. Long-term municipal bonds were paying 15 percent. I put most of our savings in those bonds, knowing that the rates would eventually go down to the normal 5-to 6-percent rate. I would collect tax-free income on our municipals, while eventually tripling the value of my bonds when the country recovered and interest rates resumed the normal range. It was a sure thing unless the country collapsed, and even Jimmy Carter could not do that. I also began investing some of our surplus dollars into oil and gas exploration, which had very high tax write-offs. That worked well to reduce our marginal tax rate, but the investments turned out to have a very low rate of return, so it took a long time to recover our capital investment. I learned not to make investments primarily to obtain tax benefits. They usually are risky and pay off very slowly. This was good experience, and Sabra and I were feeling like robber barons.

The most important study was the Continuity of Government, which included a highly classified embedded study on the survivability of the Presidency. The second study was the capability of NATO and the Warsaw Pact to mobilize for a conventional war. I was selected, because I already had a long record of possessing special security clearances and possibly because of the report I had written as a result of my participation in the Nuclear War Exercise. It must have been provided to higher levels above the Department of Defense.

For the first time in a long time, I worked only 40 hours per week with 11 federal holidays. Sabra and I enjoyed the Washington area like we never had before. We explored the Smithsonian museums, went to stage plays and concerts on the Potomac and ate in wonderful restaurants. Washington was a very inspiring and wonderful place, if you did not work in the Pentagon!! I had an underground parking space in the high-rise building above the subway terminal in Rosslyn and a corner office on the twentieth floor. Sabra would often drive up the Mount Vernon Parkway to Rosslyn, park under my building and we would take subway to a nifty restaurant or theater for an evening of entertainment. The commute was trivial. Our life style was pleasing and rewarding.

I completed both studies within two years. I had used a large number of very important consultants to review the enormous amount of data required from all 26 agencies of the federal government to assess the capability of NATO to mobilize. I was shocked that most of the 26 agencies had a lack of planning anything, other than the data required to satisfy the congressional budget inputs. We had the miraculous example of our record mobilization for WWII, which supplied the war material and weapon systems to support the U.S., Britain, and the Soviet Union. Tens of thousands of aircraft and tanks and many hundreds of ships were produced and supported almost completely by the U.S.A., since most of our allied industry was damaged. Our study found, as expected, that our country had lost that capability. We lacked the tool-and-die makers and skilled machinists that we had in the early 1940s. Our capability to do a high volume of large forgings was inadequate, and our modern weapon systems all required sophisticated alloys of scarce metals and minerals that were mined largely in Africa; e.g., titanium, beryllium, manganese. Our ground rules were to assume that you have unlimited funds, and the highest priority was how soon could you double your current maximum production rate. The answers were more than four years for aircraft engines and more than five years for Tanks and ships. The message was clear: You would fight a conventional war with what you have, and then nuclear becomes your only option. I ran the NATO capability. A CIA person ran the Warsaw Pact team, and I would guess that given the economic problems within the Soviet Union that we could outproduce them. We were not provided the results of their study, but I had read a lot about the Soviet economy and I considered it their greatest weakness.

So, in my opinion, we had lost our mobilization advantage and would fight largely with our current inventory. They had a significant advantage over us in weapon-system inventory, but we had a more favorable kill ratio with higher technology and better training. In Korea we killed 10 Migs for every F-86 lost. Congress had always squeezed us on spares and munitions inventories, and that could be the deciding factor regarding which side would have to unleash the unthinkable nuclear genie. My bosses turned the report over to President Carter's National Security staff, and I never received any feedback.

The Continuity of Government and Survival of the Presidency study was not ignored. General Brent Scowcroft, one of the President's National Security Staff advisors, took a personal interest in the Survival of the Presidency study and personally supervised the condensed briefing, which was to be briefed up the chain of command to the President. General Scowcroft was a retired USAF general who had been very effective in dealing with the Politicians. He took much of the detail out of the briefing and expressed our findings in broad policy political language, but he clearly described the threat. The briefing worked its way up to President Carter. He had already been defeated by Ronald Reagan but he strongly recommended that President Reagan and his cabinet receive the briefing as soon as possible. I felt very good.

President Reagan received the briefing and directed that all of his cabinet receive it. His Secretary of State, General Alexander Haig, had a deputy take the briefing, but all other cabinet officials were briefed. The day after Secretary Haig was to be briefed, President Reagan was shot and Secretary Haig uttered his incorrect statement, "1 am in charge." It was embarrassing but not fatal. As a result of this study, President Reagan started a multi-billion-dollar program that assured that our country would always have a President constitutionally authorized to react to an attack on the homeland.

The continuity of operations concept was enforced at the federal level and slowly spread to the state and local areas. Now, 27 years later, businesses and local governments are preparing these plans to cope with natural disasters such as hurricanes and floods.

I had enjoyed my work and comrades at TASC and had met several well-connected people, but I really wanted to leave government and go into industry. I began to circulate my resume. President Reagan was in office and the country was recovering, and once again I had hope for our country. Before I left, Four-Star General Slay joined TASC as a consultant. We had been involved together occasionally in many programs during my tour in AFLC and we shared an office at TASC. He was assigned the task of doing a study on our Critical Metals and Minerals and in a remarkably short time produced a 200+ page report on the subject that was brilliant. I was considered very productive but he was at least three times more productive than anybody in this think tank. We were probably the most politically conservative analysts in the TASC Washington organization. We both left after our studies were completed. I was destined to work for General Slay at a later date.

I interviewed for a job as executive vice president of one of the oldest banks in Washington, D.C. I would have the trust department and supervise a staff of lawyers and accountants and entertain the clients. The offer included a company car and membership in two exclusive clubs. Sabra would really enjoy this and, we would meet some very interesting people from many of the old District of Columbia families. We were enthusiastic and the bank president was arranging our contract. I received a phone call from an old friend in Honeywell whom I had worked with many years before. They were looking for a person to run the Military Avionics Division in Clearwater, Florida. I was very familiar with their principal product, an inertial platform using the electrically suspended gyroscope that had been developed in the Avionics laboratory in one of my programs in the late 1960s. It was the most accurate inertial platform for military aircraft and was installed in the B-52 strategic bomber and the new stealthy F-117. Glen Peters was the only one who interviewed me and he flew Sabra and I down to Clearwater to show me the plant. They had a single mature product, and growth applications were not promising. After our interview, he offered me the job. Very foolishly I accepted, despite the fact that the bank job was new and different and would be a clean break from the aerospace business. I called the bank president and informed him of my decision.

Honeywell flew Sabra, Kellie, and me to Clearwater to find a home and paid all expenses while we looked. Sabra chose a fully furnished model home in Countryside. We went back to Virginia and sold our home for a very nice profit, and Kellie and I returned to Clearwater and rented a condo on Clearwater Beach until we could move into our new home. Sabra managed our move, and it was the most luxurious move we had made. Honeywell treated their executives much better than the USAF treated their Generals. All of our expenses were reimbursed by Honeywell until we moved into our home. My new mortgage loan was 16 percent interest, and Honeywell made up the difference between my 7-percent mortgage rate in Virginia, paying the federal income tax withheld on the difference. Kellie enrolled in St Petersburg Junior College and I went to work.

My new organization was not a team. We had a very strong functional organization, and each function had an executive equal to my rank in Minneapolis. I was recognized as the local straw boss and the only Honeywell executive on site. All of the product program managers worked for me and were functionally supported by specialists from the functional areas. We had Design Engineering and Production Engineering, and they were not communicating adequately. There were no logistics engineering skills, and the logistics personnel had come from field support backgrounds, so they were poorly staffed to design minimizing life-cycle costs. I remembered how hard it was for them to transition from the development models into production 10 years earlier. They all were shocked that Glen Peters had selected an outsider who had never served in the Honeywell mecca of Minneapolis. I had to make these functional stovepipes into a team.

I went to Minneapolis every week for the staff meetings and discussed my concerns with my boss, Glen Peters. He was aware of the frictions between the functional areas and would support me but wanted me to keep him informed. His boss was not a strong person. He had recently been promoted into his job and had never had a chance to interview me. I could tell that he considered me an outsider. If any company needed outsiders, it was Honeywell. They talked a lot about participative management but practiced permissive management. The commercial avionics division had made a management decision to buy into the inertial platform market using ring-laser gyros, which had been developed in the Air Force Avionics Lab while I was there. They had started production and installation in airliners when I arrived and had already lost more than $100 million. I had a contract to develop and install a similar inertial platform in the US Navy Harrier aircraft. The ring-laser gyros were produced in Minneapolis, and I had to depend on them for engineering assistance; however, they were over burdened by the losses on the commercial aircraft contract. As a result they were way behind on providing needed engineering assistance to my program manager. The Navy was very upset because our schedule was slipping and very little progress was evident. I needed Glen Peters to step in and elevate this problem so that my division would not lose our Navy customer. He was unable to obtain the support, and we lost our Navy contract. It was the age-old problem that you encounter when you have a subcontractor within your own group—

you must have leverage on them. Honeywell could not, or would not, pressure their own division to provide the promised resources to my division. When you have a problem with a subcontractor, you call his boss's boss, and if you do not obtain action, you go further up the chain of command. When it is within your own group, you can only apply pressure to your equivalent level executive. I had to have my boss elevate the problem. In a good-old-boy company you may or may not receive your support. I did not. And I lost the contract.

My payroll was too large. I had a single product in production and seven products in development. I was collocated with the Space Systems Division and they had a very prosperous and expanding Space Shuttle business with NASA. I requested my counterpoint executive to consider some of my employees when he needed to fill vacancies. I had previously discussed this with my functional leaders to identify positions that they could eliminate over the next few months. I had transferred about six people to the Space division at the same pay rate when I learned that my boss had been relieved and he had retired. He had provided the only support that I had received and agreed with me that there were many problems within my division. I had made some minor improvements toward building a team, but the functional bosses felt they were losing control to my program managers. That was precisely the idea. The program managers had to be the team leaders. Our customers were always complaining of our slowness to find and correct problems, and it was always traceable to the inability to rapidly bring additional functional resources to bear. I gave them two months notice and departed Honeywell without rancor. We had a bad mismatch in management styles. I spent three years with Honeywell and learned a lot about managing for profit. My profits and bonuses were good, but it was due to my one cash cow. I thought that their long-term outlook was weak and our customers were losing faith in our ability to react to problems. My new boss thought differently. Two years after I left, they began a long slide downward. They never listened to me, because I was an outsider. They needed a lot more outsiders.

President Reagan had taken charge and I knew several of his political appointees. The son of our neighbor in Germany used to babysit our children and later came to Wright Patterson AFB as a brand new lieutenant. Lieutenant Sculley became mysteriously sick and we kept him in our quarters after he got out of the hospital. He was very interested in my Avionics experience and we talked a lot and he read a lot. He wanted an Air Force career like his father and was crushed when they finally told him he had a bad case of diabetes and would be discharged with 100-percent disability. Now, many years later, he had just been appointed to be Assistant Secretary of the Army for R&D and he had an avionic problem on the Army's next generation attack helicopter. He looked me up and asked me to assist them as a consultant. I was consulting on General Slay's team and several other agencies, but I volunteered as an unpaid consultant because it was an advanced version of our DAIS system—the one I had worked so hard to get started. The USAF was developing their next generation fighter using DAIS technology, and I knew their avionics people. I took the Army's new helicopter avionics chief engineer to meet the Air Force's new fighter avionics engineer; and, after much coordination and several months, the Army basically used much of the same architecture as the Air Force and both were happy. As a result, my consulting business improved significantly.

Sabra and I had backed Jack Eckerd, of Eckerd Drug Stores and an old Army Air Corp Pilot, for Governor of Florida against Bob Graham, who won by a landslide. After the election the state legislature criticized the State Prison Industries as inefficient and totally lacking financial visibility. They authorized the privatization of the prison industry system. Jack Eckerd had run on a platform of privatizing many Florida state agencies, so Governor Graham called Jack Eckerd and suggested that he put his money where his mouth was. I received a phone call from Floyd Glisson, whom Jack Eckerd had hired as president, and he wanted to interview me for the job of Vice President of Operations. The salary was less than half my consultant income, but the challenge intrigued me. I agreed to do the start-up until we made a profit and then would return to my consulting business.

There were 59 different industries in the 30+ Florida state prisons, with about 2400 prisoners as employees being paid up to $1 per hour. They were managed by state employees who had been associated with the prison systems for many years. We could only sell our products to government agencies, and most were sold to Florida cities, counties, and the state. The books were in very poor shape. The prison wardens did not want people who lacked experience to be within the prisons, dealing with the "cons." We negotiated that we would require their orientation course for all of our new management personnel, retain all of their state employees as our employees, and provide them with management training. Those who could apply our desired business practices could retain their management positions. Our new not-for-profit company was named Prisoner Rehabilitation Industries and Diversified Enterprises, or PRIDE of Florida.

The rehabilitation aspects had long been ignored by the State, and there were no records to indicate that much effort had been made. In 1985 it cost about $18,000 a year for each prison inmate. The best study we could find on recidivism was a federal study, which estimated that more than 75 percent of the inmates released from prison were back in prison within two years of release. The real payoff in the prison industry is to train the prisoner in a valuable skill and teach him a work ethic. Hopefully, this will significantly reduce the recidivism rate and the number of career convicts. This approach had the potential to save the taxpayer a lot of money.

Very few of the 59 industries we inherited developed valuable skills. We had many thousands of acres of farmland and forests and most of our jobs were in those operations. This included all crops, livestock, livestock products (eggs, dairy, abattoir, fertilizer), forestry, and forest products (fence posts, telephone poles, lumber, wooden furniture). These industries were valuable to develop the work ethic but involved less valuable skills. We sold many of our food crops to the prisons cheaper than market prices due to our $1/hour wages but found that our cannery was so inefficient that we could buy comparable canned products in local grocery stores at retail prices cheaper than we could produce them. I brought an expert in and we tried everything to make it work, but our small plant could never beat the big canneries, even with three shifts, which we could not do inside the prison. We shut down the cannery and used it to make cleaning supplies.

We had planted sugar cane in our prison near the everglades and made several million dollars the first year, which we used to build three new industries and improve several existing industries. Our forestry business was capital intensive but provided good steady returns. We used the Florida universities and A&M colleges to provide expertise in agriculture and forestry. We had a 35-year cycle on our pine forests: Plant very dense; thin at seven years for fence posts; 14 years for telephone posts; and 21,28 and 35 years for lumber. We used our own lumber to sell pathways, posts, and structures in our county and state parks, and we had a factory making wooden furniture for government offices. These jobs were good to establish a work habit; however, very few of the agriculture and forestry jobs required skills that paid as well as drug-dealing, prostitution, and robbing banks.

We also had several manufacturing plants. We made Florida license plates, metal office furniture, printed products, cleaning supplies, fabric products, mattresses, and remanufactured tires. Our first real rehabilitation enterprise involved our tire remanufacturing (retread) plant. The head of the Florida Remanufacturing Association protested to their Florida Senator that we were unfair competition for his associates because of our $l/hour wage rate. We offered him a tour of our facilities and wanted to discuss a solution to his problem. He accepted. We picked him up in Fort Lauderdale in Jack Eckerd's Piper Cherokee and flew him to our tire remanufacturing plant. We were very unhappy with our recently started quality program and customer complaints. We lacked the expertise to manage the plant and were looking for a consultant. We could hardly wait to talk with him. He was slightly aggressive until we showed him our plant, and then he started being cheerful as he explained that our equipment was obsolete and the plant layout and everything was wrong. He had thought we could sell to commercial companies, but when he found out we were only authorized to sell to governments, he told us we were not even competitive at $l/hour. He was ready to eat lunch and go home. We took him to lunch and asked him if he would be interested in helping us become more competitive and possibly become a partner in our business. We would provide the capital to procure the needed equipment and modify the layout as he recommended. He would provide us with a plant manager who would put in a quality and financial system and run our plant. Our key interest was that he might provide jobs in one of his association's plants when our inmates were released from prison, based on the recommendation of our plant manager, whom he had provided us. In turn we would give him a percentage of the increased profit and we would have a start on our rehabilitation initiative. He called us back a week later and accepted. It would become the model for our rehabilitation efforts. The company is now, 22 years later, named Pride Enterprises, has $80 million in sales and most of their businesses are modeled even better than our first enterprise, which is still operating.

I really enjoyed this job. The prisoners were a real challenge. The young ones looked like the young men and women I had worked with for 30 years in the USAF. They were not. They could look you right in the eye and lie like a pro. The prison superintendents were worried that our new managers would be conned by the cons. We had made all new employees go through orientation, and we continued to emphasize this. Even after these precautions, I had to fire two managers who were conned into forbidden activities by the cons. My worst case was a manager conned into providing contraband to a sweet young lady who looked like Shirley Temple and provided him sex for favors. We had to look at the prisoners' records if we spent time with them, and I always queried several prisoners as I toured their workplaces. They were eager to talk and I would listen and therefore had to read their record. You have to work hard to get into a state prison. Most had long records of arrests and imprisonment in county jails.

The older inmates were quieter, and you could tell who were leaders and which ones were institutionalized. The institutionalized inmates had served many years and had grown used to prison. Often after release they would go into a store and try to rob it but not try too hard to escape. They would be sent back to prison as they intended. We had one inmate—a good worker—near the end of his sentence, and he was a trustee with a good prison record; so we trusted him to work in the warehouse outside the prison walls. He was on his last day of work with us and he turned up missing. The search found him less than two miles away out on the highway hitchhiking in his prison garb. He wanted to be caught so that he could stay in prison.

The toughest prisoners were the Cubans that Castro had released from his prison and mental institutions and transported to Florida via the Mariel boat invasion. My Cubans were in a medium-security area within the prison in Glades, Florida. We hired the best of them to work in our abattoir, killing the cattle and butchering and packaging the beef for market. They looked like they had no soul, with flat, dead, and dark eyes. I always wore a suit and my abattoir manager would always take me down the line to watch these troops carving the beef with their very sharp knives. It was the only time I felt a little insecure.

The really sad cases were the white-collar prisoners who had lived normal lives and ended up in prison for a single crime— murdering their spouse/boss/neighbor, mercy killing of a suffering terminally ill spouse, grand theft, etc. They deserved the time but it was really tough on them.

I really enjoyed the job and stayed an extra year beyond my start-up commitment. It was very challenging and I learned a little bit about a lot of industries, applying my MBA degree to the best of my ability. I had done all of the financial analysis on each industry and provided my managers with a fund-flow diagram so they could have a feel for the finances and also gave them graphics on their fixed and variable cost so they could see how products should be priced. If their break-even price was below the market price, we took a closer look; and, if we could not make it work, we closed the business and used the facility for another business. I closed five businesses and started four new ones. They finally bought a simple financial management system after I left, but we worked off of books for my two years with the company. I reluctantly departed after hiring my replacement. He stayed with the company until he retired. My consulting contacts welcomed me back from my brief stay in the Florida prisons.

CHAPTER 17

Consulting, and Enjoying Life: 1985-1997

Sabra and I decided we had more than enough income to cover our own expenses, with our children out of college. It was time to do what we wanted to do and enjoy life. When I retired from the USAF, my wife received a card from an old dear friend with the message: "Congratulations on your husband's retirement. Half as much income and twice as much husband!" We laughed but it was true. We were only 55 but decided I would consult on jobs I enjoyed but not become tied down to the point where we were not free to take long breaks for cruises or just weekends in neat places. Our marriage and family life had been wonderful, but the next 17 years were even better. Sabra was beautiful and could still wear her college clothes. Our love seemed to grow stronger with time, and that was because I tried to build my faith to be as strong as Sabra's. Everyone could sense her soul. I was amazed that she did not glow in the dark.

Tina and Kevin were married, and Kellie had graduated from the University of South Florida. After six years in four colleges, Kellie followed us from job to job. Tina and Brian were in the US Navy and had blessed us with a wonderful granddaughter and

grandson and we visited them often. It was great to have babies again. Kevin was starting out as a health insurance agent under the wing of an older agent but owned his own condo in Columbus, Ohio. A little starvation is good for the soul. Kellie was living with us and working for a very grumpy lawyer as his paralegal. One day my unpredictable daughter came home and told me she had enlisted in the US Army for two years as a private. She had a college degree and enlisted as a private because she did not want to commit for a longer time. Sabra and I prayed for her while she was in boot camp. I wish I could have witnessed her reaction to her drill sergeant. After boot camp she was assigned to a heavy equipment repair company, and my five- foot, three-inch daughter was driving really big road-building equipment. Sabra and I visited her in Fort Leonard Woods, Missouri. Her Sergeant told us they called her "Mother Superior," and she would allow no foul language in her presence. He claimed that to his amazement it worked! Of course she had been promoted to Private First Class, so it was only to be expected that my tiny daughter could reform an entire Army company. The good news was that she enrolled in night graduate school and earned a Master's Degree in Human Resources Management. She also met red-headed First Lieu-

Jay and Kellie's wedding: Bob, Sabra, Kellie, Jay, Tina, Kevin, granddaughter Sabra in front

tenant Jay Hicks in one of her graduate classes and later married him. Kellie spent many weekends with my mother and Aunt Gladys in their retirement condos in St. Louis, Missouri. She loved both of them and would stay up late eating popcorn in bed and telling them about her Army life and her new boy friend. All of our children believed their grandmothers and Aunt Gladys were supreme beings, and they were well loved by all of their relatives.

Munich Hofbräuhaus 1989

Kellie served out her two years in the Army and returned home to us. Jay and Kellie were married in our Catholic Parish in Safety Harbor and we had their reception at the local country club. She was our last child to leave the nest and we were alone at last in our home in Clearwater, Florida. We would see every one of our children at least once a year for the rest of our lives and generally spend vacations with them either in Pensacola, Florida, or with my sister and her family up in the northern peninsula in Michigan. We also exchanged visits with my mother and Aunt Gladys frequently. Both Sabra and I were very family-oriented and believed in the three (or more) generation family, although it was becoming more difficult in our geographically dispersed family units.

I began consulting with the Institute of Defense Analysis (IDA) in the system analysis division. I worked every other week at IDA in Arlington, Virginia, and stayed in the Radisson Hotel in Alexandria. I negotiated a luxury room on the concierge level, guaranteeing 10 days per month, and Sabra joined me frequently. We knew the manager, the concierge, and all of the room clerks and maids. When Sabra was with me, they would give us one of the honeymoon suites and we would write a long critique of the room for their enjoyment. The honeymoon suites were very plush and romantic with four-poster beds, heart-shape saunas, mirrors, and soft lighting. At the end of our honeymoon week we would complain that the lighting was to dim to read in bed, or the television was badly corroded and looked like it had been broken for a long time, or the foot bath (bidette) was not shaped like a foot, or some other nonsense, which the staff enjoyed. In the evening we ate out at our favorite restaurants, visited with former neighbors, and enjoyed the Washington area. I worked on projects requested by Congress or the Office of the Secretary of Defense; e.g., the ability of the B-52, B-1B, and the B-2 to penetrate the Soviet Union (3 studies) and the effectiveness of the resistance to electronic jamming of various avionic subsystems. After two years I was becoming too busy with other consulting jobs and stopped working with IDA. The other jobs were more lucrative and interesting. Sabra and I missed our visits to Washington.

I had a few consulting jobs on my own, but most of them were as an associate of General Slay in his Slay Enterprise Inc. He had contracts with all of the large aerospace companies, and we began to specialize in evaluating company proposals, responding to Government Requests For Proposals on major weapon systems. At one point he had a team of about 60 retired military and civilian associates that he could draw upon, tailored to satisfy the expertise required by that specific weapon system. I believe we had the highest win record of anyone in this business, and I would guess it was better than 85 percent. The contracts were very large, ranging from several million up to several hundred billion dollars. We were mostly on Department of Defense contracts, but we also won several contracts with Westinghouse for nuclear reactors.

I had worked and flown with many members of our consulting team during my military career. Most of us had retired as Generals or Admirals. I had flown F-86s with Lt/General Tom McMullen when we were Lieutenants in Korea in 1953. I had worked with and been friends with General Al Slay, Lt/Generals Al Casey and Truman Spangrud, Major/Generals Don Litke, Ted Broadwater, and Dick Steere, and Colonel Bill Crabtree since the 1970s. Our consulting group and our wives had an annual symposium every year in a nice hotel somewhere in the United States and enjoyed the festivities very much. Working and relaxing with these lifelong friends is one of our fondest memories.

Our biggest and most enjoyable win was the USAF and USN Advanced Tactical Fighter. We were on the Lockheed/Boeing/General Dynamics team against the Northrop Grumman/McDonald Douglas Team. We worked on that proposal for more than four years. The proposal was 75 volumes and more than 35,000 pages. The contract was estimated to be more than 400 billion dollars. Our prototype aircraft won the fly-off in 1993, and the first Squadron of F-22s became operational in 2005. It embodied everything that our Digital Avionics Information System Laboratory team had envisioned in 1968 and more. In

the 22 years with General Slay I worked more than 200 contracts. I specialized in digital avionics, systems engineering, software, human factors, weapon guidance, and risk management. I worked less than half of his contracts because Sabra and I needed time to enjoy the fruits of our labor.

We had kept our model home—the one we had bought fully furnished in the Countryside subdivision in Clearwater, Florida. It was our best home ever with a screened-in swimming pool, four bedrooms, three and one-half baths, and a large beautiful yard. We had a lot of friendly neighbors and a friendly Catholic church with a child-friendly Irish priest. Sabra and I belonged to the Safety Harbor Spa and worked out there three to four times a week. There were several county parks nearby that we walked frequently, and we were less than five miles from Clearwater Beach and Honeymoon Island off the coast from Dunedin. We joined a Yacht Club, where you paid an initial fee and then could call with a day's notice for a fully equipped boat up to 28 feet long, and it would be ready and waiting for you and your guests to climb in and cruise off. When you returned, you filled it with gas, paid your reasonable rental rate and gas bill, and left with your guest. The Club lasted for about three years and never made it a success. By this time we knew that we certainly did not want to own a boat, because all of our friends who did moaned and groaned about what a good deal we had to avoid the frequent maintenance, clean-up and tender loving care required of boat owners. We enjoyed our brief love affair with small boats without the burden of ownership. Then we started on big boats ... usually called cruise ships!

Our first and best cruise was to Alaska, where I had worked as a boy on three sealing expeditions and experienced my one and only really, really cold-weather flying as an eager Lieutenant. We talked Ray and Mel Grazier into going with us and we signed up with Princess Cruise Lines for a 15 day trip—eight days on land and seven days on the inland waterway, which I had traversed six times as an enthusiastic boy. This was our best cruise ever. We stayed two nights in the Princess Lodge in the Kenai Peninsula, where we went

Whitewater rafting and hiking. Then we took a full-day train ride in a Princess dining car with a plexiglass roof through the Denali National Park to the Princess Lodge in the middle of Denali. We stayed three nights there with one full day on a bus, climbing up a mountain from the dense forest up to above the tree line to the top of the mountain where we had a picnic. The top of the mountain was a religious experience, with only the sound of the wind and the majestic eagles. We could see more than 100 miles in every direction, and it was all nature without any sign of man or man's creations. We saw a lot of wildlife on the way up and down the mountain, with grizzly bears stealing the show. We left Denali on the train for another full-day trip through the park with its magnificent preview of nature. Now we were in Fairbanks in a hotel with one day on a riverboat, cruising the Yukon River. Then we had an airboat ride up a very shallow and rapid river to a gold mining operation and fur trading camp out in the wilderness. After that, we flew to Anchorage and took a bus back to our ship at Seward for the seven-day cruise back to Vancouver.

The cruise back through the inland waterway brought back many memories. The highlight was Glacier Bay, where we anchored about one-half mile from the glacier for a whole day, listening to the low rumble of the glacier as it very slowly flowed down between the mountains and into the sea. The low frequency rumble was sporadic, and the calving of the glacier as large chunks fell into the sea mesmerized us. The waves from the ice chunks came out to the ship in ever-expanding concentric circles. We stopped in Sitka, Juneau, and Ketchican, and they were all crowded with cruise ships and tourists—a far cry from the days when our Seal Blubberers Barbershop Quartet stole the show in the local bar and brothel. The inland passageway cruise was spectacular as usual. The only shock was the smaller number of whales and Orcas compared to our sealing voyages fifty years earlier. This was probably caused by the huge Japanese factory ships that slaughter the whales at sea. Once again, it was an enormous adventure.

We took Tina's children, Sabra and Brandon, on a Disney cruise when they were 10 and 12 years old. Disney used the Celebrity cruise lines out of Canaveral Florida before he built his own ship. The four of us had a big cabin and had a ball. The grandchildren enjoyed fun and games with their own age group and really enjoyed it. I was sure that Brandon would fall overboard. He was running everywhere at breakneck speed. Somehow he survived and Sabra

Northern Peninsula - Mom, my family, and Doris & Phil's family plus Dawn & Dog

and I had another joyful experience to store in our memory bank.

We had frequent family get-togethers at our home in Clearwater, at Doris and Phil's cabin in the Northern Peninsula in Michigan, or on Pensacola Beach. Mom, Gladys, my family, Doris and Phil's family, and Sabra's sisters and brother were the usual participants. Swimming, boating, game-playing and hiking

Bob and Sabra with grandchildren Brandon, Bobby, Jayson and Sabra

**My family, Phil and Doris's family,
Ben and Molly's family and David Bickerstaff's family**

prevailed.

Every time Kellie's (Army) or Tina's (Navy) husband was transferred to a new base, we would visit them after they settled down and explore the local attractions with them and our grandchildren. Both families had 22-year military careers. With Tina it was Oceana, Virginia; Key West, Florida; Albuquerque, New Mexico; Pensacola, Florida; Lemoore, California; and Jacksonville, Florida. Water sports, swimming, sailing, fishing and seafood dominated our time, along with hiking in the Redwood Forest and touring San Francisco and the wine country. With Kellie's family it was Fort Leonard Wood, Missouri; Aberdeen, Maryland; Giessen, Germany; Ft. Sill, Oklahoma; Boise, Idaho; Ft. Hood, Texas; Kaiserslautern, Germany; and Tampa, Florida. We shared many adventures in Europe with them, including Rhine and Moselle River cruises, trips to Paris, Austria and Venice, wine-tasing, and hiking in several national parks in the United States.

We were with Kellie's family in the Lodge in Yellowstone National Park when Sabra's sister, Molly, called us and told us that her husband, Ben, had been killed when his private plane crashed. He was flying alone and apparently became unconscious and crashed into a farm field. It was a terrible tragedy, and Molly and her three daughters were devastated. Their strong faith carried them through this ordeal. Ben was a wonderful husband and father, and they had invested wisely enough to provide his family with life-long security. My prayers for him continue.

Our travel adventures continued. We took Mom and Gladys down to Patrick AFB Florida to watch a night launch of the Space Shuttle. We stayed in the General Officers four-bedroom quarters on the beach across the bay from the launch pad. I awoke them at 4 a.m., wrapped them in blankets, and took them out to the beach to watch the liftoff. It was astounding—we could feel the sound waves and sense the power of the engines as the shuttle zoomed off into space. Mom and Gladys had already experienced the birth of autos and airplanes, but this stole the show. They were exhilarated!

Between cruises, we took a lot of side trips within the United States. We had built up a large number of frequent flyer and hotel points and would use them to visit relatives or points of interest. We spent a long weekend in New Orleans, where Sabra had gone to college at St. Mary's Dominican, stayed in the French Quarter and had a ball.

We bought front-row-center tickets for three Broadway shows nine months in advance, booked first-class round-trip tickets and five nights in the Marriot Marquis Hotel at Times Square and saw Cats, Phantom of the Opera and Annie. We toured Manhattan in double-deck buses and ferry boats and walked in Central Park. Mayor Giuliani had already cleaned up New York. I would never have taken Sabra there the way it was five years earlier. It was fun and exciting.

We were very proud of the fact that our son, Kevin, had won a trip for two to Ireland from one of the health insurance companies that he represented. He was one of their highest volume insurance agents. We were proud that he was awarded this honor, but Sabra was overjoyed when he selected to take her as his guest. (I stayed home and disowned him.) They toured Ireland, operating out of a luxury hotel in Shannon and then one in Dublin. They saw most of the attractions, ate the good Irish food, and hit a few of the pubs. Kevin still talks about the science involved in drawing a Guiness draft beer. Sabra talked about it for the rest of her life. All of her children loved her, and she loved them and our grandchildren unconditionally, but this was a very special treat from her son just for her. She never forgot it.

We visited my mother's recently found cousin (on my grandfather's side) in Concord, Massachusetts. Bob and Mary Ann Clark were researching her family tree and had found and fell in love with my mother and her sister, my Aunt Gladys. Bob was a talented computer expert and could find work anywhere, so they had made their life an adventure, living and working in strange exotic places. They had rented Ralph Waldo Emerson's home in Concord, Massachusetts. They were the first family to live in this enormous mansion who were not direct descendants of Mr. Emerson. They begged Mother and us to visit, and we were very glad that we did. The house was very old and had many bedrooms, all with fireplaces. The first evening, they invited for cocktails Colonel Emerson, USAF Retired, a direct descendant of Ralph Waldo Emerson, to meet us. As we were talking, the door to the library slowly opened ... and then slowly closed. We all stopped talking and Colonel Emerson explained that was the Reverend Dutton, the family ghost. As was the custom in those days, when the local reverend retired he was taken in as a family member by one of the local wealthy families. Ralph Waldo Emerson had taken in Reverend Dutton, who stayed with the family for many years and died in the house that had become his home. Colonel Emerson explained that the reverend was harmless and never had disrupted the family. Bob and Mary Ann's son and their dog had encountered Reverend Dutton and both were afraid of him, but it was hard for us to believe him, or Colonel Emerson. That night Mom did not sleep a wink because she felt that someone was in her room. I thought it was the power of suggestion that had caused her to be nervous, but Bob and Mary Ann moved her to another of the many bedrooms. The next night I went up to her new bedroom to light the fire in the fireplace. As I bent over to light the fire, I definitely knew somebody or something was behind me. If I'd had hair on my spine, it would have been standing straight up—a very primal reaction, which I never have experienced before or since. I told Sabra (but nobody else), and we slept that night about one-eighth of an inch apart. Mom slept peacefully all night. I believed in that ghost and felt that he resented the presence of a strange family in his home. In the next few days we walked all around Concord, following the path of the first battle of the Revolutionary War and through the Harvard Woods and then took the train into Boston and people-watched. We had a lot of fun and good food, and enjoyed everything but the Most Reverend Dutton.

We were very fortunate to have a good honest family gentleman, Mike Bilirakis, as our representative in the U. S. House of Representatives. Sabra and I and many of our friends worked to help his initial and several subsequent election campaigns. After he was elected I volunteered to lead his committee to interview and rank-order the candidates from his district seeking scholarships to the USAF Academy. Our committee of four to six people would interview up to 35 candidates and rank-order the top ten by overall score as his nominees. This made them eligible for academy appointments. Then we would have our Academy liaison officer and our committee members work our list often nominees seeking athletic or superintendent appointments. When all possibilities were exhausted, Congressman Bilirakis would use his appointment on our highest ranking nominee not yet selected. We generally would end up with two or thee of our ten nominees entering the USAF Academy, and one year we enabled five to attend. During the 15 years that I served, I was very impressed by the high quality of all of the young people that we nominated. It was very satisfying to me to know that our fixture USAF would someday be led by these excellent young leaders. Congressman Bilirakis would always sponsor a banquet for all of his USAF, US Army, US Navy, U.S. Coast Guard and the Merchant Marine Academy appointees and their parents, which Sabra and I always attended. I am certain that some of these young men and women would make General and most would make full Colonel. It still gives me a good feeling.

We were stationed together with Dan and Carolyn Cheatham several times in our career. It seemed like they would be assigned to another base and a few months later we would be assigned to the same base. Our son, Kevin, was the same age as one of their sons, Ray. They had all boys, and Kevin played over there most of the time because his sisters always had girls at our home—the Cheathams had the "boys club." So Ray and Kevin became inseparable from grade school up into their working careers, and all of Ray's brothers loved Kevin. Tragically, one night Ray and Kevin were driving up a narrow road and Ray was passing Kevin's car on his motorcycle when they came to a bridge. Ray hit the left edge of the bridge and was killed. Both families were devastated, and Kevin grieved for Ray for many years. I still remember Ray in my prayers.

Sabra and I had been occasionally visiting my cousin Bee who was living down in Fort Lauderdale by herself. Her husband had died several years earlier and her mother, my wild aunt Bess, had died earlier. She had agoraphobia for years, was dying of emphysema and still smoked. Her children lived out west and were in poor financial shape, and she was being cared for by an elderly friend. She had been robbed twice and her neighborhood was not safe. Sabra and I decided to fly her in an air ambulance up to a nearby nursing home. She weighed 85 pounds and would probably die within a few months. I had always been close to Bee and Gordon and their two children. I was a board member for a 300-unit life-care retirement facility for retired military officers and their widows and had good contacts in the nursing home community. I knew she would receive good care in this facility, which was run by a former nun. She had the first 90 days under Medicare and probably would not last that long. I went through all of her personal papers and she had less than $50,000. Then I found a copy of her father's trust. He had died more than 25 years earlier. I contacted the Bank in St. Louis and they still had more than $400,000 in the trust and had not heard from the family for years, so the funds had kept growing. Further, the trust would cover all expenses for health care. I brought them up to date with death certificates and names and addresses of her children and made arrangements for the nursing home to send the bank all bills. Sabra and I visited Bee every week, or more if she called. She had to go on 100 percent oxygen and stop smoking, so I kept her in butterscotch candies to substitute for her cigarettes. She lived for two and one-half years in the nursing home, fighting for every breath. It was a horrible way to die, and we had to leave before she died because Sabra's mother died at age 92. The nursing home was wonderful, but Sabra and I bought a long-term care insurance policy and vowed we would never place each other in a nursing home. I split up Bee's remaining assets and sent them to her children. They also received the remainder of the trust courtesy of their grandfather, one of my two early male role models, Seney Huckstep.

One day my sister, Doris, called Mom and me and told us she had cancer of the kidney and they were removing the kidney. Mom went up to Ann Arbor,

My sister, Doris, and her wonderful husband, Phil Jackson

Doris and I after Phil died

Michigan, to stay with her after the operation. Several months later, on her first checkup, they found cancer in her other kidney, her lungs, and her pancreas. Her doctor could do nothing more, but Doris would not give up. She had worked in the hospital and kept bugging the doctor to put her in a drug-testing program. He finally found a program that would accept terminally ill cancer patients. The drug was artificial interpheron alpha, a chemical that replicated this ingredient in the human immune system. Miraculously, the drug and Doris' positive attitude cleared up the cancer in her kidney and pancreas and reduced the cancer in her lungs. The lung cancer kept recurring, but she lived for five more years, long enough to see two more grandchildren born. Her husband, Phil, my male role model, died suddenly of a massive heart attack while playing tennis with his best friend. Doris was determined to beat her cancer, and I believe that was at least half of the battle; however, when the love of her life died, she gave up the battle and the cancer came back in spades. She died one year after her husband's death. Her children, grandchildren, Mom, Sabra and I were left to grieve the loss of both of them. First Sabra's mother, then Bee, then Phil, and now my big sister, who was dead at age 65.

Tina with Sabra and Brandon

Aunt Gladys' husband, David Bond, died and Mom and Gladys were alone in their nice retirement condo in downtown St. Louis. Their brother, Donald Williamson, had also lived in the same condo and he also died. Now they were alone in St. Louis and all of their friends and family were gone. Gladys was over 90 and wanted to give up driving. They both were in very good health and Mom still walked every day; however, downtown St. Louis was risky. Sabra and I talked it over and agreed they should come down to Clearwater and live with us. We discussed it with them. They were very reluctant to live with us but agreed to move to Clearwater. I found a very nice retirement facility just two miles from our home, and we took all of the literature and drove up to St. Louis to help them pack. While we were there, we drove around the west end of St. Louis where we had lived before we left. It was now a decimated and dangerous neighborhood but it still renewed old memories of our big family and wonderful neighbors. Sabra took Mom and Gladys in our very comfortable Lincoln Town Car and I drove Gladys' 1959 Plymouth (with less than 60,000 miles) back to Clearwater. They stayed with us until they made up their minds to have individual units in the new retirement home. They were worried that they would outlive their life savings, but I assured them that I could find investments that would more than make their life savings adequate. I opened joint accounts with both of them and our investments provided more than enough income to keep them very comfortable. They both received monthly statements from Vanguard, so they always knew their financial status. Both of them died with more savings than they had arrived with. They moved into their new homes and were very disoriented for about a week, which upset Sabra and me. The manager assured us that this was normal and would soon pass. Within two weeks they were playing bridge every day, had made friends with many people and told us this was much better than St. Louis. We were very relieved and within a month their social schedule was full and Mom had a boyfriend. We usually took them to the shopping center for lunch on Friday and ate dinner with them in their excellent dining room once a week. Occasionally, we would go on side trips to nearby tourist traps.

When I was not consulting, we played golf several times a week with Betty and Bill Ketelhut and enjoyed it very much. I shot in the middle 90s and Sabra was about 105. Bill was a little better than I and Sabra a little better than Betty. We took several weekend trips to Florida and Georgia golf Courses. We enjoyed walking, bridge, and golf together until the Ketelhuts moved to Orlando in 1997.

Jim and Valerie Kennedy's Wedding in Belfast, Northern Ireland

We heard from my half brother Jim Kennedy in 1995. He and his new wife, Valerie, flew over from Northern Ireland to visit us in Clearwater.

We had not been in contact since March 15, 1980, when they informed me that our father had died. They met Sabra, Mom, and Gladys and made a big hit with my family. They had just married after a 30-year courtship. Their mothers had both died, and our older brother and sister had also died, so both of them were without family. We happily adopted them.

Mom meets her new adopted son, Jim Kennedy

Jim and our sister, by coincidence, had purchased a second home in Homosassa, Florida, in 1992 just 54 miles up the highway from Clear-

Bob and Sabra visit Jim and Valerie in their home in Northern Ireland

water and stayed there occasionally during the Irish winters. They visited during our best-weather month and it rained every single day they were with us! We returned their visit to Northern Ireland the next spring during their rainy season and it was fair and sunny every day, reaching a very unusual 90 degrees temperature. It was clear that Jim brought rain and I brought sunshine! My mother loved him immediately.

They immigrated to the U. S. two years later with a visa and lived in Homosassa, and we saw them frequently. We all visited Pensacola Beach on vacation together with our children and grandchildren and met many of Sabra's family in Pensacola. They were now fully integrated into our family. When I introduced Jim to our close friends as my brother, they raised their eyebrows because they never knew I had a brother and we did not look alike, similar to Arnold Schwarzenegger and Danny DeVito. We explained that we were identical twins, separated at birth, and recently reunited. That seemed to satisfy their curiosity.

We had a couples' social group at our church, which we all enjoyed. Sabra and I were among the youngest. Over the years the men began dying, and before long there were only three men left. We plugged along, though, until bad health began to reduce our numbers below two tables. Then we degenerated down to one Superbowl party per year. We had a good time together while it lasted.

CHAPTER 18

Finale: 1997-2002

Sabra and I, and Bill and Betty Ketelhut, took a ten-day cruise to the western Mediterranean, beginning in Barcelona, Spain, and ending in Dover, England. It was a first-class experience on a Cunard ship. We stopped in Casablanca, Morroco; a small port in Spain; Lisbon, Portugal; Bilbao, Spain; two small ports in France; and then sailed on to Dover, England. It was spectacular but about half as good as Alaska. We took a train from Dover to London and stayed a few days, sightseeing, before going home to Clearwater.

We also took two trips in paddlewheel riverboats. Our first was a five-day trip up and down the Mississippi River from New Orleans. The second was viewing the fall colors from Pittsburg down the Ohio River to Columbus, Ohio, where we spent some time with Kevin and his wife. The riverboats were a lot of fun with most of the advantages of larger cruise ships but with stops at good old American small towns on the river.

Sabra had two cataract operations and emerged with double vision. She had to live her remaining life with very thick corrective eyeglasses, but her vision was saved. Shortly after that, Mom, Gladys, Sabra and I were having lunch in an old hotel in St. Petersburg, Florida, and Sabra suddenly went blind in her right eye. We called her eye surgeon, and he told us to come right over and to minimize Sabra's motions. It was a detached retina, and he operated the next morning to move it back in place. She regained her vision and again corrective lenses restored it adequately.

Mom, Gladys, Sabra and I were having our normal Friday lunch in our favorite Scnickelfritz deli, when Mom fell and broke her hip. She was 92 and her

Our Mississippi Queen River Trip

surgeon marveled that her she was in such good shape. The operation pinned her hip together, but she was unconscious for four hours. She was rehabilitated in a nursing home and was their star pupil, but she had lost some degree of alertness, probably due to the anesthesia. One year later they had to repin her hip, and again she was unconscious for several hours. She never was the same again. She had delusions of little people in her home who would not pay rent, and some dementia. She stayed in her retirement facility in the assisted care unit, and I would bring her to our home weekends or whenever she felt endangered by her uninvited guests. One day they found her unconscious on the floor. She was taken to the hospital and they placed her in hospice care. She was dying. Sabra and I stayed with her until she died. We both were with her holding her hands when she regained consciousness and took her hand away from me to gesture up to the ceiling rising up staring happily at whatever vision she had on the ceiling. The nurse said she had seen that many times and liked to think that she saw the angels coming for her. Sabra and I decided that was true. She died that night. I held a memorial service for her in her retirement home with all of her grandchildren, Gladys, Sabra, and all of our Clearwater friends. I buried her ashes next to her mother and wonderful father in Oak Grove Cemetery in St. Louis, Missouri. Uncle Dale Williamson, David Bond, and Uncle Seney Huckstep are in the same plot, and Uncle Don Williamson is buried nearby. Just right for one of their spirited Pinochle games that I remember so well from my listening post under the dining room table.

I inherited Mom's Vanguard account as the remaining joint tenant. I knew that Aunt Gladys had supported Mom, Doris, and me after my grandfather had died, and that she had given Mom $25,000 before they moved down to Clearwater. She also had loaned Doris $20,000, and Doris had died before she could repay her. Thus the first $45,000 went to my aunt Gladys. I believe my family were the only ones that ever repaid her for her most generous contributions made during her lifetime. She had bought me my only bicycle. She was very touched when I acknowledged her contributions, and there were tears in her eyes. I split up the remaining money among the six grandchildren.

Most of the family had never accepted Gladys' second husband, David Bond, because he was dating her before his first wife died and some ugly episodes occurred. This hurt her very much that her own family could not forgive them. She had lived with Mom most of her life, and we wanted to take care of her as the last survivor of my family and my co-mother.

Mom, Gladys, and I had the same family doctor, Dr. Mermelstein, a wonderful and caring young Jewish doctor. He loved Mom and Gladys and was very proud of his oldest patient. I always took them to the doctor and monitored their visits. Gladys had always had a cholesterol count above 300 and Dr. Mermelstein could not talk her into taking medication to reduce the reading. He was very proud after her 101st birthday and took her hand and respectfully requested if she would fill and take his prescription. Her response will never be forgotten in the annals of medical history. Aunt Gladys took his hand with both of her hands, looked gently into his eyes, and said, "My cholesterol has been this high most of my life. I am 101 years old. It's a bullet that already missed." She passed the rest of her physical with flying colors. I flew off to a consulting job in Los Angeles, and two days later Sabra called me and told me that Gladys had called her, said she felt bad and asked to be taken to the hospital. Gladys hated to go to doctors or hospitals, so that worried both of us. Sabra and my sister's daughter, Mary Jackson, took her to the hospital and stayed with her until visiting hours were over. The next morning the hospital took her down to the laboratory to start a series of tests. She died on the table without a word before the testing started. The doctor said her heart and lungs just stopped functioning. Her time had come. No pain, but she died alone. We held another memorial service in the retirement home with all of our family and David Bonds' family. Aunt Gladys was cremated and buried in her family plot at Oak Grove in St. Louis, next to her husband and my mom, whom she had lived with most of her life.

I had been worried about Sabra for several years. She had lost her ability to plan and execute parties, meals, trips and things that were normally second nature to her. Kellie confronted me and expressed her concern. I listened but did not respond. Kevin and Tina also mentioned their concern after my mom's death. I had been packing Sabra's suitcase for cruises and trips for several years. The episodes became more and more frequent, and finally one day she took a walk by herself and could not find her way home. I think that this had started more than five years earlier, and neither of us wanted to admit to reality. We had raised our family and were millionaires, and just wanted to enjoy the rest of our life together. I took her to her doctor, and after several tests he diagnosed her with Alzheimer's disease with Parkinson's symptoms. He had read recently about this combination and referred her to the brain center at the University of South Florida. After many visits and tests, they diagnosed her with Alzheimer's and probably Lewy Body Disease, recently discovered by neurologists in England. Once diagnosed, the degradation accelerated. Denial may have been a better alternative. I vowed to keep her at home and care for her myself. She had pills for depression, pills for dementia, pills for hallucinations, and pills for Alzheimer's, and I had to give them to her. I doubted if the pills worked because her depression, hallucinations, and dementia just galloped along.

Tina's husband of 22 years had asked her for a divorce. He was still in the Navy and had worked his way up through the enlisted rates, earned his way into college, and was commissioned as an officer. After her divorce, Tina moved home to help me take care of Sabra. Her daughter Sabra (named for her grandmother) stayed in Oklahoma City to marry her high school sweetheart and started a career in restaurant management.

Our son, Kevin, had met and fallen in love with Vira, a lovely Indonesian lady, whom he had met in Columbus, Ohio. She had graduated from Ohio State University and was employed as the Chief Financial Officer of a small steel company. They were married in a gala three-day ceremony in her hometown in Medan, Indonesia, honeymooning in Bali. It was a beautiful ceremony, and they now reside in Powell, Ohio. Sabra and I were unable to attend the ceremony, but I know Sabra is happy for them.

I decided to move to Pensacola Beach, where we had courted 50 years earlier. I could not take care of our big home and Sabra too, and we had more family in Pensacola. We bought a nice condo on Pensacola Beach as a temporary home, while a more luxurious condo was being built. Tina rented a home in nearby Gulf Breeze and continued to help me with Sabra. I could no longer handle her—my back was becoming strained, because I did not know how to handle an impaired person. I activated our long-term care policy and hired a nurse assistant to care for her six hours a day. I finally found an excellent young nurse who was very good with Sabra. She could handle her in the bath, help her dress, and was very patient with her. Best of all, they would talk and laugh and both felt at home with each other.

Kevin & Vira's wedding in Indonesia

The Bride and Groom's Dance

Our restaurant trips grew fewer. The last time we went, I took her to the ladies' room halfway through the meal and when we returned someone else was seated at our table. I stopped taking her to Sunday mass because she could not make it through the whole mass. Her cousin, Father Thorsen, came and gave us mass in the condo. Our long walks on the beach became shorter and shorter, and she began falling occasionally, due to her Parkinson's disease. We finally settled on having picnic lunches in one of the Fort Pickens beach shelters right on the Gulf of Mexico. I fixed lunch while her nurse dressed her, and the three of us would drive two miles up Fort Pickens Road to our beach shelter, out of sight from everybody, and sit in the shade, eat lunch and listen to the Gulf. We still slept in the same bed and cuddled before we went to sleep, and we both needed all the affection that we could cram into these precious moments. Her decline seemed to be slowing down, and I hoped that she would last until we moved into our luxury condo.

One day I received a phone call from a bank inquiring about Sabra's application for a credit card. I advised the caller that she was ill and could not have applied for the card. He put his supervisor on, and we arranged for him to send me the application after I faxed him my power of attorney for Sabra. This happened one more time, and now I had two filled-in applications with her name and social security number and two different addresses. I called our Escambia County sheriffs office and talked to their deputy who handled identity theft. He reviewed the data and stated that no money had been stolen yet, and identity theft could not be proved with the evidence that I had provided. I would have to gather more evidence while he sat in his office eating doughnuts. I was furious with this redneck law officer but did not want the incompetent Escambia County Sheriffs office involved. I contacted the postal inspector in Tallahassee, and he drove 200 miles to investigate this case. He froze all mail addressed to Sabra's name at those two addresses until both applications had been received, and then he photographed the people who picked up the letters. The postal inspector then received both applications that had been mailed to the banks, and he had his case. I was heartbroken—it was Sabra's nurse and her new boyfriend. They could not make a case against the boyfriend, and our nurse pleaded guilty. She wrote me a tearful letter with an apology. I believe the boyfriend put her up to it. I wrote a letter to the judge and requested that she be given probation, and he sentenced her to two years and released her on probation. I never told Sabra. I hired a new nurse; and, after a few days, Sabra liked her too.

Sabra's decline hastened and she was not always certain of my identity. Her only brother, Phillip Joe, had been diagnosed with Alzheimer's following very serious heart surgery. He died suddenly of a heart attack before his Alzheimer's became serious. I told her, and it did not seem to register, and then I knew that most of her memory was gone. We did not attend the funeral because they were very close and I was glad that she remained unaware of his tragic death.

Several weeks later Sabra had a really bad night. When the nurse arrived, her blood pressure and pulse were high and she could not talk. The nurse called 911 and the doctor in the hospital verified a stroke. When she emerged from intensive care, they placed her in a hospice room in the hospital. They expected her to die but could not say how long. Her cousin, Father Thorsen, gave her the Last Rites at her bedside. Kellie and Kevin had flown in, and we stayed with her day and night. I slept in a bed beside her and held her hand, but she was not fully conscious. After a week, the neurologist said the stroke was not that serious but her brain was badly impaired by the Lewy Body disease so a little stroke damage goes a long way. She said I could take Sabra home with hospice care and that she might last weeks or months. Kellie's family was stationed in Germany, and Kevin had an insurance agency to run, so I sent them home. Hospice set up a hospital bed and some medical equipment in my condo, and we took Sabra home in an ambulance. The hospice nurse could not stay with her, as I had thought, so I brought back my hired nurse for day duty with Tina, and I took the night duty. She had a terrible night and when the nurse showed up at 6 a.m., she said she was dying. Tina called Sabra's sisters, and the hospice nurse instructed me how to keep her throat passage clear and how to give her morphine if she was in pain. My hired nurse did not have the credentials to perform these tasks. The hospice nurse then left and gave me her cell phone number ... to call when Sabra died!! This was totally different from the hospice people with my sister Doris, Bee, and Mother, but this was Escambia County. Sabra died at 10 a.m. surrounded by her sisters, Tina, and me. I talked to her until I knew she was gone. The hospice nurse returned at noon and officially declared her dead. We had been very happily married for 49 years, five months, and eight days.

Kellie and Kevin returned for the funeral. I picked out a beautiful black evening dress with a single rose embroidered on it, and had a hair stylist fix her hair. Her pearl earrings with a small rose were her only jewelry, and she looked like her glamorous self. We had an excellent funeral director. All of Sabra's sisters, nieces, and nephews came from far and near, as well as all of our close friends in the area. We had a viewing night followed by the rosary. The next morning we had her funeral mass in St. Ann's Church in Gulf Breeze, with Father Thorsen assisting Monsignor Hunt. After the mass, a very long funeral procession took us to Barrancas National Cemetery on NAS Pensacola for a graveside service and burial. Her tombstone was engraved: "Beautiful soul, wife and mother." I will be buried alongside her someday, and our souls will once again be rejoined as they were many years before. I miss her very much but she remains in our memories forever.

Memories

As the years flew by our family grew. First Mary Kellie, then Kevin Carter and finally Katherine Martina joined the family. Sabra was a very active and loving mother and encouraged all of the neighborhood children to play in our home or yard. She even took some of them along on family vacations. At the children's school she assisted teachers by tutoring students in math and reading. As a Camp Fire Girl's leader she took her group on overnight camping trips, day hikes in the woods and swimming in nearby rivers and lakes. Throughout the years Sabra continued to teach her children and grandchildren and their friends and set an example for them to follow while loving and enjoying them. Sabra was also active in the Officers Wives Club, assisting in many charitable endeavors. Later she sponsored the Debutante's formal ball and summer coming-out events for four years at Wright Patterson Air Force Base, teaching young college girls to be a lady.

Sabra has made many homes for her family during her 49 year marriage: Laredo, Texas; Las Vegas, Nevada; Fort Walton Beach, Florida; Montgomery, Alabama (3 times); Stillwater, Oklahoma; Kaiserslautern, Germany; Ramstein, AB Germany; Wiesbaden, Germany; Dayton, Ohio(4 times); Alexandria, Virginia (2 times); Clearwater, Florida and now finally Pensacola Beach, Florida. When hurricanes came, I left her and the children to save airplanes and when snowstorms came I could not return home until the runway was cleared. One time she was snowbound for three days with three teenagers and three cars that would not start. I had more than two years separation from my family for overseas tours in Korea and Vietnam and countless short separations for military temporary duty or business trips. Mary Sabra coped with it all.

Sabra was a good Catholic with a deep faith. She sent her children to Catholic Schools and raised them as Catholics. I was not a Catholic but took instructions, studied and tried hard to acquire the faith because I admired the strength and grace in Sabra's faith. One of the many priests that gave me instructions summarized my problem. Like doubting Thomas, I seemed to want to witness a miracle, but as was often pointed out, true faith requires no proof. One day I received orders for an F-105D combat tour where the combat losses were very high. It was time for a decision. After 14 years of waffling I promptly converted, successfully tested my new faith on my combat tour and returned home as a practicing Catholic. My conversion only made a good marriage even better.

Sabra loved people, particularly children. She had fun with children and truly enjoyed them. She particularly loved her grandchildren; Sabra, Brandon, Jayson, and Bobby. She loved her family and the people whom she grew up with in Pensacola. She loved her military friends, neighbors and their children and grandchildren, whom she met during our 30 year military career and 22 year civilian career. She has been well loved by all friends and family.

Sabra was adored by my entire family and neither of us ever had any in-law problems. My doting grandmother, mother, sister and aunts consistently sided with Sabra and enjoyed her grace and goodness. Sabra provided loving companionship and later cared for my mother and aunt until they died in 1998 and 2000 at the ages of 94 and 101.

Sabra loved the arts and our many museums. She frequently took her family and the neighborhood kids to the Smithsonian Museums in Washington. She enjoyed the Kennedy Center, front row-center seats for Broadway shows in New York, the Getty Museums in Bilbao, Spain and California, the Air Force Museum in Dayton, and local museums and theatre wherever we lived or traveled. In Pensacola she has enjoyed the pure white sand on Pensacola Beach, the Saenger Theatre, the historic district, the Naval Air Museum, the new art stores downtown, Trader Johns, Fort Pickens, the National Seashore, the cliffs and beach along Scenic Highway and Fort Barrancas.

Sabra loved God's earth and travel. She lived four years in Germany and skied in Germany, Austria and Zermat, Switzerland, drank beer and sang in the Munich Beer Hall, cruised the Rhein and Moselle rivers, enjoyed many wine fests and hiked in the Black Forest. She hiked or cruised in many of our public and National parks from Denali and Glacier Bay in Alaska, to Yellowstone, Giant Redwoods, Grand Canyon, the Petrified Forest and Glen Helen near Yellow Springs, Ohio. Our family once spent a long weekend in Germany flying both ways in Air Force 2, the Vice President's aircraft. The family traveled down the Miami River in canoes in Ohio, enjoyed Whitewater rafting in the Kenai Peninsula in Alaska and camped out in Prince William Park in Virginia (where our neighborhood kids proved to be better winter hiker/campers than the local scouts). She has cruised the lower Mississippi River, the Ohio River, the Yukon River, the inland waterway from Vancouver to Alaska, the Bahamas with her first two grandchildren, the Eastern Mediterranean, Straight of Gibraltar and the English Channel with ports of call in Spain, Morocco, Portugal, France, and England. She has toured Paris, France; Venice, Italy and Northern Ireland as a guest of her half brother-in-law, Ireland as a guest of her son, and Puerto Rico and Mexico. In later years she enjoyed General Slay's annual Symposiums across the United States and the camaraderie within our senior aerospace consultant group, some of whom we have known for 50 years.

Now she has returned home to her first love; Pensacola and her very large, very loving family and friends, to complete the full and happy circle of her life.

It was not so many years ago
In this kingdom by the sea
That there lived a couple whom you may know
Bob and Sabra Kennedy;
And this couple, they lived with no other thought
Than to love and be loved by thee.
Now this couple lay in Barrancas
Fanned by a breeze from the sea
O'erhead the roar of jet engines
That keep our dear country free.
From the home of the Navy's Blue Angels
Our Angel will rise up to thee.

—Adapted from Annabelle Lee by Edgar Allen Poe

The original poem was the first poem I had to memorize and recite in Miss Gray's 4th grade class at Dozier Grade School in St. Louis, Missouri.

This is our Prayer, developed over our lifetime:

Dear God, Heavenly Father, we thank you very much for the gifts that you have bestowed upon us. First, we thank you for the opportunity for an immortal soul. May our souls be joined in heaven with those whom we love. Second, we thank you for the diversity of our peoples and their cultures. May we adopt the best practices of these diverse cultures into one and learn to live together in peace.

We thank you for our planet Earth, with sufficient food, water, and shelter to support all of your people, and more than enough beauty to inspire our creativity. We thank you for our universe with all of its miracles and unknowns, sufficient to challenge our minds until the end of time. Finally, we thank you for your Son and all of your saints and prophets whom you have sent to guide us to your way of life and our place in life. For all of these gifts, we thank you and praise your name. We ask that you bless our souls, the souls of our families and friends, and the souls of our descendants and ancestors, with faith, strength, health, happiness in life... and togetherness in heaven,

Through Christ our lord. Amen.

Sabra & Bob

22 December 1952 - 1 June 2002

Until death do us part...

Made in the USA
Monee, IL
04 November 2026

67387239R00157